# Power of Nature

This book presents a compelling exploration of the five fundamental elements—soil, air, heat, water, and space—that constitute all animate and inanimate entities on Earth. Drawing from ancient philosophical traditions across Indian, Chinese, and Greek civilizations, this groundbreaking work establishes a comprehensive framework for understanding the elemental basis of existence. The text illuminates heat as the most critical element sustaining all living species, while examining humanity's perpetual struggle to comprehend and mitigate natural disasters like earthquakes, typhoons, and volcanic eruptions. Through meticulous analysis, the author navigates the paradox of using the mind—itself perceived as the supreme controller of all species' activities—to understand the very nature that created it. Central to this work is Nature's remarkable capacity for transformation, a unique feature enabling its power potential for productive applications across various domains. The interdependence of all five elements is established as essential for survival, with the absence of any single element rendering life impossible. The book also introduces an innovative classification system for species based on sensory capabilities, ranging from one-sense to six-sense organisms, providing a new taxonomic perspective on biodiversity. This volume will be of great interest to scholars of philosophy of science, environmental philosophy, ecological sciences, science and technology studies, and comparative religious studies. It offers valuable insights for researchers exploring traditional knowledge systems, sustainability practitioners, and graduate students in interdisciplinary environmental studies, environmental engineering, botany, and zoology seeking to bridge ancient wisdom with contemporary ecological challenges.

**Ethirajan Rathakrishnan** is Professor of Aerospace Engineering, IIT Kanpur, India. *Rathakrishnan Limit* and the *Breathing Blunt Nose* are his contributions to technology. He is Fellow of the Royal Aeronautical Society. He is editor-in-chief of the *International Review of Aerospace Engineering* and *International Review of Mechanical Engineering*. He has authored many books, including *Mind Power: The Sixth Sense*, Routledge, 2023.

# Power of Nature

Ethirajan Rathakrishnan

Routledge
Taylor & Francis Group
LONDON AND NEW YORK

First published 2026
by Routledge
4 Park Square, Milton Park, Abingdon, Oxon OX14 4RN

and by Routledge
605 Third Avenue, New York, NY 10158

*Routledge is an imprint of the Taylor & Francis Group, an informa business*



*British Library Cataloguing-in-Publication Data*
A catalogue record for this book is available from the British Library

ISBN: 978-1-041-25085-2 (hbk)
ISBN: 978-1-041-25087-6 (pbk)
ISBN: 978-1-003-74505-1 (ebk)

DOI: 10.4324/9781003745051

Typeset in Sabon
by Apex CoVantage, LLC

This book is dedicated to my parents,
*Mr. Thammanur Shunmugam Ethirajan*
*and*
*Mrs. Aandaal Ethirajan*

# Contents

# Preface

This book aims to address the issue associated with Nature, which is the power potential or energy source that creates, nurtures, and terminates all the animate and inanimate species that prevail in the universe.

Soil, water, fire, air, and space are the five elements that constitute life on the Earth. All animate and inanimate species on the Earth are formed with different combinations of these basic elements. Now, the confusion that might come to our mind is: Who is the creator or controller of these formations? This book focuses on addressing this specific question.

It is expected that this book will be of some value to the readers in getting an insight into the secrets, power, and control of Nature's potential for the benefit of the world.

Ethirajan Rathakrishnan

# 1 Introduction

Nature is the power potential or energy source that creates, nurtures, and terminates all the animate and inanimate species that prevail in the universe. It is well known that all living species, beginning from one-sense plant kingdom to six-sense humans, are made up of the five basic elements of Nature, namely soil, air, water, heat, and space.[1]

*Soil:* Represents solid matter, the ground, and the physical world.

*Air:* Represents motion, breath, and the intangible nature of wind.

*Water:* Refers to liquid substances, fluids, and the life-giving properties of water.

*Heat:* Symbolizes energy and its transformation.

*Space:* Represents emptiness, void, and the vastness of the universe.

These elements are the foundation of many ancient philosophies and world views, including those in Indian, Chinese, and Greek traditions. Among these, heat is envisaged as the most vital element that makes the life of all living species. This perception may be because the body of all living species is warmer when alive, and becomes cold when dead. This perception is due to the scientific feature that views energy as heat.[2] Science proves that heat energy can be used to generate momentum that can be used for work. However, in this conversion, science experiences that, in principle, the total amount of the energy of a system cannot be converted to useful work. This leads to the development of another representation known as the *enthalpy* for energy. Enthalpy is the sum of the useful form of energy known as *kinetic energy* and the form that is not yet understood by science, termed as *internal energy*. Science continuously strives to scale to the potential of Nature. This effort of science to estimate the quantum of energy shows that it is impossible to quantify the energy potential. This may be because the mental energy (the capacity of the mind to perform tasks like problem-solving, decision-making,

DOI: 10.4324/9781003745051-1

and maintaining attention) required to scale this has to be derived from the intuitive power of the mind governed by Nature itself. This kind of peculiar entanglement makes the scaling energy content of Nature a Herculean task.

With the marginal understanding gained by science, let us venture into gaining more understanding of Nature. Before entering into this process, let us try to consolidate the knowledge gained by the scientific world on the potential of Nature. It is almost true for all branches of science that the awareness about the need to understand Nature erupted from the natural calamities faced by humans. For instance, natural disasters such as earthquakes, typhoons, and volcanic eruptions threaten the human race whenever they happen. This fear paved the path for probing into these events to protect them from their fury.

If we analyse the developments to protect life from earthquakes, the first and foremost idea that might have come to the human mind is to make shelters that can withstand the fury of the violent jolts caused by the movement of the plates of the Earth. Once it was understood that earthquakes are due to the movement of two or more plates, the process of developing technology to build houses with basements that would move sideways, carrying the entire building as a single unit so that there would not be any damage to the house structure. This thought process resulted in the development of technology to construct earthquake-resistant buildings in places such as Japan that is prone to frequent earthquakes. The testimony to the success of the human mind in the development of this technology is the safety provided by the buildings built with this technique during the severe earthquake faced by Fukushima city of Japan on March 11, 2011. This was one of the worst earthquakes the world has ever faced. The intensity of this jolt was 9.7/10 on the Richter scale. For earthquake intensity measurement the Richter magnitude scale (often shortened to the Richter scale) is the most common unit for measurement of earthquakes.

A closer look at the power of the earthquake and the consequences associated with that power will be of value to gain an understanding of the energy potential of Nature. An earthquake has immense energy to shake the Earth's mass of huge magnitude, which cannot be moved even to an insignificant length, to many centimetres sideways, or even up and down in some cases. For example, the Fukushima earthquake on March 11, 2011, that lasted for about 20 minutes was quite unusual not only in its time duration but also in the shaking style. In addition to lasting for about 20 minutes, it shook the Earth sideways in the first 10 minutes, and up and down, as a horse gallop, in the second 10 minutes.

History reveals that most scientific developments are need-based. When humans lived in the Stone Age, there was no need for cultivation in the beginning, as they were accustomed to natural food such as fruits, roots, fish, and meat. Initially, humans ate these as raw. Maybe when they had tasted the meat of some animals burnt in a forest fire, they might have felt the importance of cooking the meat to improve its taste. This might have initiated the use of fire for human use. As the population increased, there was a need to go some distance in search of food items, such as fish, meat, and roots. During the root collection, they might have tasted some seeds such as grass seeds. This might have been the beginning of the use of food grains, such as rice and wheat.

As time passed, they might have felt the need to produce the grain to meet their demand by cultivating the food grains. This might have made them think about the need for cultivation tools such as spades, ploughs, and knives. This process might have initiated the metal age, leading to the making of iron metal from iron ore. This could have ignited the metal age in human history. The basic elements involved in the process of making iron are the Earth (iron ore as the raw material) and fire (coal as the fuel). Once the cultivation begins, there is a need for soil, water, heat, air, and space. Thus, for any production, all five basic elements are required. However, the proportion in which these elements should be used for productive processes varies from process to process.

One question that keeps lingering in the human mind is which is the most powerful among the five basic elements that constitute the universe. Though the world strives hard to find an answer to this fascinating doubt, it has yet to find an answer. The main barrier that prevents the world from finding an answer to this doubt is the fury of these elements, during their abnormal behaviour, experienced by the world. For example, the Earth is so patient with all human activities—and sometimes atrocities—without showing its fury in the form of seismic behaviour. Sometimes seismic activity results in tsunamis. The magnitude, energy release, and shaking intensity are all related measurements of an earthquake that are often confused with one another. Their dependencies and relationships can be complicated, and even one of these components alone can be confusing. One way to measure the magnitude of an earthquake is to compute the energy released by it. The amount of energy radiated by an earthquake is a measure of the potential for the magnitude of the damage it can cause to man-made structures. An earthquake releases energy at many frequencies; therefore, to compute the energy released by it, all frequencies of shaking for the entire event are to be accounted for. However, as we can visualize, this is not easy. This kind of complication makes the quantification of the energy content of a seismic activity complex.

Despite this complexity, the scientific world not only continues to quantify the power of Nature but also strives to overpower that. With this understanding, let us get into the Herculean task of gaining an understanding of Nature without getting into the rigorous scientific theories. Our aim here is to gain an insight into the power of Nature from the natural events taking place in the life of the living species on Mother Earth.

# 2 Basic Elements of Nature

We saw that soil, water, fire, air, and space are the five elements that constitute life on the Earth. All animate and inanimate species on the Earth are formed with different combinations of these basic elements. Now, the confusion that might come to our mind is: Who is the creator or controller of these formations? Indeed, from time immemorial, the scientific and spiritual worlds have kept thinking about this to get an affirmative answer to this fundamental question. However, despite this continuous effort to find a definite answer to this secrecy, by science and theism, there has been no success in this venture, until now. Let us try to view the hurdles that prevent the understanding of the secrecies of Nature.

In the view of the author, the first and foremost obstacle to the understanding of Nature is that the wisdom required for this Herculean task has to come from the mind, which is perceived as the supreme controller of all the activities of not only humans but also other species. The pathos at this stage is that the power required for the functioning of the mind has to come from the body in which the mind is embedded. It is known that the body itself is a device made up of the five basic elements of Nature. Thus, the power required for a mind to understand Nature has to come from Nature itself. This requirement imposes the stipulation that the mind that attempts to understand the power of the secrecy embedded in Nature has to learn to control and focus the thought waves emitted by it on the problem to be solved. This learning that enables a person to approach a problem with focus is termed 'single-minded devotion' by the scientific world and meditation by the spiritual world. To achieve this state leading to the attainment of the wisdom required to understand the power of Nature, one has to study the problem to be understood, without any perturbation to the mind waves, and focus them only on the problem, like a concave lens making the sun rays converge on a single point. In simple language, this approach may be stated as something identical to a laser beam that propagates without dispersion.

Only a mind capable of emitting wisdom waves of a single frequency and uniform amplitude, like white noise, would be successful in gaining an understanding of the power potential of Nature. One of the main features that

DOI: 10.4324/9781003745051-2

make Nature manipulate its power potential for productive use is its capability to change from one form to another. This feature makes Nature unique in all respects. In other words, it may be stated that the ability of Nature to create species that possess one or more forms of energy embedded in it and any one or all of these energy constituents can be released when subjected to a suitable environment. For instance, the thermal energy possessed by firewood is liberated when ignited.

The capability of Nature to change its composition from one form to another makes it suitable for productive activities. For example, the volcanic lava and ash erupting from active volcanoes enrich the soil of the Earth, making it suitable for rice cultivation. A typical example of this is the ash constantly ejected by the active volcano on Sakurajima Island adjacent to Kagoshima City in the southern part of Japan.[3] The volcanic ash makes the soil so rich to yield turnip—a round root with white or creamy flush and edible leaves, which is considered a vegetable—as big as weighing about 40 kg each. Another simple example is the water in the tender coconut. Even in the coastal area, the coconut tree roots draw and send the water to the treetop and convert it to sweet water, which is unique by the tree, and stored inside the coconut shell. These several events can be cited for the ability of Nature to change from one form to another in all kinds of organic and inorganic items.

Some of the popular events personifying the transforming power of Nature are the germination of seeds and the change of carbon to diamond. A seed contains life cells, vitamins, and minerals, which are in a kind of hibernation until it encounters moisture for its germination activity.[4] The moment it encounters moisture, the life cells sleeping inside begin their recreation activity and grow as a plant of their parental nature, gaining enough strength to tear off the cover or shell, irrespective of the cover's strength, and come out. Now the baby plant meets the environment containing air, heat, and space required for its further growth, and the soil needed to firmly fix its main root and root branches. This friendly environment enables the plant to grow steadily, passing its main root, penetrating deep, and the surrounding roots spreading sideways.

Similarly, all species on Earth grow with the help of the five basic elements of Nature. Among these five elements, if one element ceases to exit the species cannot survive. For example, on planets such as the moon, there is only soil, heat, and space but the air and water are absent.[5] Because of the absence of two basic elements, there is no life on the moon. This implies that all five elements of Nature are essential for the existence of life. Also, it is not just their existence but they must also be in proper proportion with each other at the required level for the life of a particular species. For instance, some species living in water cannot survive when kept out of water. Similarly, many species living on land cannot survive in water.

# 3 Classification of Species

All species on Earth are made up of the five basic elements of Nature. However, the senses possessed by these species are not the same. Based on the number of senses they possess, the species can broadly be classified into six categories. They are the one-sense, two-sense, three-sense, four-sense, five-sense, and six-sense species. Tamil grammar book *Tholkappiam*[1] lists these categories. The species of the plant kingdom are one-sense species, such as trees, plants, and grass, having only sense to feel the touch. Species such as snails and shells are two-sense species that can sense touch and taste. Ants and millipedes, which can sense touch, taste, and smell, are three-sense species. Crabs and beetles, which have senses of touch, taste, smell, and vision, are four-sense species. Birds and animals having the senses of touch, taste, smell, vision, and hearing are five-sense species. Humans are the only six-sense species having the ability to feel touch, taste, smell, vision, hear, and infer with analytics. Though the above classification, which states that only humans have the sixth sense, namely the logical sense of inferring with analytics is widely accepted, a recent book, *Mind Power: The Sixth Sense*[6] not only questions this special quality attributed only to humans but also convincingly establishes that the intuitive thinking capability is common to all species, irrespective of the number of senses the species are capable of.

Now the question that comes to our mind is this: Does it mean that the previous theories about the species classification and the inference that humans are the only species created by Nature with six senses and the rest of the species are inferior to humans? Come on, let us strive hard to find a convincing answer to this involved, but genuine, question.

Before getting into this Herculean task, let us understand what is meant by rational thinking, that is something special, and humans are the only living species in Nature blessed with that. The understanding of the world about this is the ability to analyse something and infer the message conveyed by that. For instance, if we feel severe heat on a summer day, we infer that the heat content in the atmosphere is very high, and that is due to the sunshine. Indeed, the sun can be regarded as the power source for the life of all species living on Earth. In other words, it may be stated that for the evolution and

DOI: 10.4324/9781003745051-3

living of all species, beginning from one-sense species to six-sense species the energy source required is provided by the sun and there would not be life on Earth without sun.

Most of the sun's energy reaching the Earth includes visible light and infrared radiation but some is in the form of plasma and solar wind particles. Other forms of radiation from the sun can reach Earth as part of the solar wind, but in smaller quantities and with longer travel times. The sun is a hot sphere of gas whose internal temperatures reach over 20 million Kelvin due to nuclear fusion reactions at the sun's core which convert hydrogen to helium. According to Cool Cosmos—Caltech, 'The Sun's interior is primarily composed of hydrogen (approximately 70%) and helium (approximately 28%)'. The rest is made up of smaller amounts of other elements like oxygen, carbon, neon, and iron. These elements exist in a state of plasma due to the extreme heat and pressure within the sun's core.[7]

The radiation from the inner core is not visible as it is strongly absorbed by a layer of hydrogen atoms closer to the sun's surface. Heat is transferred through this layer by convection.[8] The surface of the sun, called the photosphere, has a temperature of about 6,000 K and closely approximates a blackbody. For simplicity, the 6,000 K spectrum is commonly used in detailed balance calculations but temperatures of 5,762 ± 50 K[9] and 5,730 ± 90 K[10] have also been proposed as a more accurate fit to the sun's spectrum. Astronomers use 5778 K when classifying the sun as a star.

The sun is the star at the centre of the Solar System. It is a massive, hot ball of plasma, inflated and heated by energy produced by nuclear fusion reactions at its core. Part of this internal energy is emitted from its surface as light, ultraviolet, and infrared radiation, providing most of the energy for life on Earth. Its diameter is about 1,391,400 km. The sun has been an object of veneration throughout human history. In many cultures, the sun was thought to be a solar deity or other supernatural entity. It has been a central subject for astronomical research since ancient times. The sun has extremely important influences on our planet: It drives weather, ocean currents, seasons, and climate, and makes plant life possible through photosynthesis.

All life on Earth is supported by the sun. The sun radiates energy and provides us with both heat and light by fusing hydrogen into helium at its core. We call this as solar radiation. Only about half of this solar radiation makes it to the Earth's surface. The rest is either absorbed or reflected by clouds and the atmosphere. Still, we receive enough power from the sun to meet the power demands of mankind and other species living on Earth. Solar energy—power from the sun—is a vast, inexhaustible, and clean resource. At present, scientists from many countries are working on solar panels to convert solar energy to useful forms, such as electric current as efficiently as possible. Indeed, this effort is successful to a great extent.

# 4 Species–Nature Relationship

The human–Nature relationship goes beyond the extent to which an individual believes or feels they are part of Nature. It can also be understood as, and inclusive of, our adaptive synergy with Nature as well as our long-standing actions and experiences that connect us to Nature. From time immemorial, humans and other species have learnt to adjust to Nature for their survival. The simple fact that most of the species continue to live is testimony to the capability of the species to adjust to the changes in Nature. It is well known that lions usually live in places with jungle having bushes and plains, and not with dense forest cover. This enables them to hunt for their food comfortably. However, there are exceptions to this general behaviour of lions. Lions live in deserts also.[11] The only reason we can assign to this strange behaviour of lions living in drastically different environments is that time and the natural environment will train a species to adjust to the conditions that prevail in their living place and enable a species to survive. In other words, it may also be stated that if a species cannot adjust to its surroundings it cannot survive. Historians speak about the species that extinguished is the evidence for this fact. Several books explore the topic of extinct species. Elizabeth Kolbert[12] examines the current mass extinction event and its causes.

Another interesting feature about the Nature the scientists find is that there are many new species, in all the species living on Earth, that keep surfacing periodically owing to cross-pollination. However, this cross-pollination is confined to the respective categories only and does not cross this boundary. For example, crossing between two animal species is possible, but not between animal and bird. This kind of regulation in the species' evolution might be regarded as the effect due to the Nature discipline. If Nature is so disciplined, then why does it show its fury such as earthquakes, tsunamis, cyclones, hurricanes, typhoons, and volcanic eruptions? Scientists strive hard to find answers for this harsh behaviour of Nature. Also, they try to quantify the power/energy associated with these calamities.

In the deep ocean, a tsunami can move as fast as a jet plane, over 500 mph (805 km/hr), and its wavelength, the distance from crest to crest, maybe hundreds of miles.[13] Once a tsunami has been generated, its energy is distributed throughout the water column, regardless of the ocean's depth. A

DOI: 10.4324/9781003745051-4

tsunami is made up of a series of very long waves. The waves will travel outwards on the surface of the ocean in all directions away from the source area, much like the ripples caused by throwing a rock into a pond. The wavelength of the tsunami waves and their period will depend on the generating mechanism and the dimensions of the source event. If the tsunami is generated from a large earthquake over a large area, its initial wavelength and period will be greater. If the tsunami is caused by a local landslide, then both its initial wavelength and period will be shorter. The period of the tsunami waves may range from 5 to 90 minutes. The wave crests of a tsunami can range from a few to a hundred kilometres or more apart as they travel across the ocean. As the waves approach the coast, their wavelength decreases and wave height increases.

On the open ocean, the wavelength of a tsunami, the horizontal distance between two corresponding points on consecutive waves, typically measured from crest to crest, maybe as much as 200 kilometres, many times greater than the ocean depth, which is of the order of a few kilometres. In the deep ocean, the height of the tsunami from trough to crest may be only a few centimetres to a metre or more—again depending on the generating source. Tsunami waves in the deep ocean can travel at high speeds for long periods for distances of thousands of kilometres and lose very little energy in the process. The deeper the water, the greater the speed of tsunami waves. For example, at the deepest ocean depths, the tsunami wave speed can be as much as 800 km/hr—about the same as that of a jet aircraft. Because the average depth of the Pacific Ocean is about 4000 m (14,000 feet), tsunami wave speed will average about 200 m/s or over 700 km/hr (430 mph). At such high speeds, a tsunami generated in the Aleutian Islands may reach Hawaii in less than four and a half hours. In 1960, great tsunami waves generated in Chile reached Japan, more than 16,800 km away in less than 24 hours, killing hundreds of people.[14] The power of a tsunami comes from the following. An earthquake suddenly pushes part of the seafloor up or down, which changes the height of the water above it—what physicists call potential energy—and the potential energy quickly changes into the kinetic energy of the tsunami waves.

In a rough guess, Harry Yeh, a professor of ocean engineering at Oregon State University, said that the earthquake on March 11, 2011, at Fukushima coast in Japan, pushed a section of seafloor 250 miles long and 50 miles down by an average of one yard, resulting in billions of cubic yards of water—trillions of pounds—suddenly shifting position.[15] That energy going into the tsunami, according to Professor Yeh's estimate,[15] was a bit less than that of an exploding atomic bomb. In addition to the damage that a tsunami can inflict along coastlines in particular countries, it can also affect the entire Earth. The planet's oceans are very heavy, applying enormous pressure to the ocean crust. When the distribution of that pressure is shifted, as it is during an earthquake, it can induce wobbles in the Earth's rotation.

It takes a lot of force just to keep water in place. More than three million cubic yards of concrete hold back the water behind Hoover Dam on the

Nevada–Arizona border. At its base, the pressure of the water reaches 45,000 pounds per square foot. In that case, however, the enormous power of water is put to positive use: to generate electricity.[15]

The power of an earthquake is expressed in terms of magnitude on a scale called the Richter scale. An earthquake of 2.0 or less magnitude on the Richter scale can be felt only a little. An earthquake over 5.0 magnitude on a Richter scale can cause damage from things falling. An earthquake with 6.0 or higher intensity magnitude is considered very strong and 7.0 is classified as a major earthquake. Earthquakes occur when elastic energy is accumulated slowly within the Earth's crust as a result of plate motions and then released suddenly at fractures in the crust called faults. The released energy travels in the form of waves called seismic waves.

The energy release can also be roughly estimated by converting the moment magnitude, $M_w$, to energy using the equation

$$\log E = 5.24 + 1.44\, M_w$$

where $M_w$ is the moment magnitude. An earthquake intensity of 6/10 on the Richter scale releases $7.8 \times 10^{13}$ joules. This amounts to $29 \times 10^6$ hp. The energy of this order is released by Nature. Now, the question that arises in our mind is, where does this energy come from or where is this quantum of energy embedded in Nature? The answer to this curious question is that no one knows the origin of this energy released by Nature.

Volcanic eruption is another natural phenomenon that has immense thermal energy.[16] Ninety-nine percent of the gas molecules emitted during a volcanic eruption are water vapour ($H_2O$), carbon dioxide ($CO_2$), and sulphur dioxide ($SO_2$). The remaining 1% is comprised of small amounts of hydrogen sulphide, carbon monoxide, hydrogen chloride, hydrogen fluoride, and other minor gas species.

Like earthquakes, volcanic eruptions, and cyclones, there are numerous phenomena in Nature, which involve a huge quantum of energy associated with them. However, until today, there is no scientific theory to explain the source of this energy. This forces scientists to state that the power of Nature is beyond the perception of the human mind.

# 5 Creative Nature

In addition to creating all animate and inanimate things in the universe, Nature is responsible for all the literary creations available to us. We can see that there is no creation without the influence of Nature, whether it is authored by theists or atheists. In ancient times, almost all countries of the world believed that natural resources such as the sun, air, soil, water, and space were in some way or other responsible for life on Earth. This faith made people worship these. A typical example of this worship is the faith by the Shinto sect in Japan. Some of the well-known sects based on Nature are pantheism [a doctrine which identifies GOD with the universe, or regards the universe as a manifestation of GOD], theism [belief in the existence of a GOD or GODs, specifically of a creator who intervenes in the universe], panentheism [the belief or doctrine that GOD is greater than the universe and includes and interpenetrates it], deism [the belief in the existence of GOD—often, but not necessarily, an impersonal and incomprehensible GOD who does not intervene in the universe after creating it, solely based on rational thought without any reliance on revealed religions or religious authority], polytheism [the belief in or worship of more than one GOD], animism [the belief in a supernatural power that organizes and animates the material universe], totemism [a belief system that establishes a symbolic connection between a group of people and a natural entity, such as an animal, plant, or even a natural phenomenon], shamanism [a spiritual practice where practitioners, known as shamans, interact with a spirit world to address various concerns in the physical world, often through altered states of consciousness like trance], paganism [a modern religious movement incorporating beliefs or practices from outside the main world religions, especially Nature worship], Saridharam [a Nature-based religion practised primarily in the eastern Indian states of Jharkhand, West Bengal, Bihar, Odisha, and Assam], Sarnaism [characterized by the worship of Nature, including trees, hills, and forests], Kirat [the name Kirat has multiple meanings and origins, primarily signifying praising GOD or singing GOD's praise], and Wicca [a religion influenced by pre-Christian beliefs and practices of western Europe that affirms the existence of supernatural power]. All these sects worship Nature in one form or another.

DOI: 10.4324/9781003745051-5

In the physical world, Nature is perceived as the source for all living beings and understanding their constituents and control are beyond the power of the human mind. This concept or doctrine appears to the religious consciousness as a manifestation of the sacred or supreme power. Through Nature, the modes of the species are quite different. It is not known whether species other than humans think of the supreme power and formulate anything like religion. The human race postulated many theories in the name of the supreme power and termed them religion. The sun, the moon, and the Earth, for example, can symbolize realities that transcend human experience. Throughout the history of religions, 'Nature' has been perceived as initiating a relationship with humankind. In large part, this relationship is expressed in forms of adoration, a response of the total personality, or an entire religious community, to the phenomena of nature.

Nature worship underscores the fact that the sacred can appear in any guise. A theist is confronted by the paradox that the sacred can manifest itself in material form without losing its basic character. In the worship of Nature, radically different levels of existence are felt to interpenetrate and coexist. The possibilities of the human spirit become coextensive with the sacred capacities of the rest of the physical universe. Thus, Nature worship highlights both the freedom of the sacred power to appear in any form and the capacity of the human being to recognize it. It also underlines the capacity of profane reality itself to become a transparent symbol of something other than itself. In such a religious perception, Nature transcends its brute physicality. It becomes a cipher, a symbol of something beyond itself. From this point of view, Nature's existence is like the human situation in the world. Its modes of being as a manifestation of the sacred power become resources for understanding the human religious condition. In many traditions, the belief in the shared destiny of Nature and humanity is highly elaborated, so that the objects of Nature are held to possess the same essential qualities as human beings: emotions, life cycles, personalities, volition, and so on.

The value and function of Nature thus go beyond the concrete sphere to the mystery of the sacred as it appears in the fuller reaches of religious experience. Only by keeping this in mind will people understand the forms in which communities respond to powers revealed in the physical universe.

The sky is often revered as a manifestation of divinity or venerated as the locus of the GODs. Objects falling from the sky come from the sacred locus of the heavens and often become the objects of religious cults. Actual meteorites are frequently the centre of a cult associated with sky GODs. In the same way, flints and other species of 'thunder stones' or 'rain stones' fallen from the sky are treated as sacred, for they are believed to be the arrow points shot by the GOD of lightning or by other celestial divinities.

Worship of the sun is widespread, especially at the times of the solstices. In societies engaged in intensive agriculture, the sun is worshipped in connection with the fertility of the crops and the regenerative life of the cosmos. Such is the case with Inti in the Inca pantheon (a complex system of deities with Viracocha as the supreme creator GOD). The sun's power in such cases is not limited to the fertility of foodstuffs but also extends to human progeny. In many cultures, the sun is believed to traverse the underworld at night.[17] Therefore, the sun becomes a sacred guide for the soul's journey through the land of the dead. Veneration of the sun takes the form of following his tracks when he sets. The sun carries into heaven the warriors who have fallen in battle. Frequently, the sun is worshipped because of its heroic achievements, including the creation of human beings. The moon is one of the most fascinating and rich religious characters. It has long been an object of worship in many cultures. The moon's shifting shape and changing disposition in the sky at various times of the night, day, and month make it the focus of a wide range of associations that have led to its veneration.

The moon is frequently a lascivious being associated with the wanton powers of fertility. Often the moon is venerated as the source of sexual life and the originator of reproductive processes, such as menstruation and intercourse.

Mountains are a ubiquitous object of cult. The mountain is a divine body in whose life all beings participate and from whose abundance and well-being all benefit. The community cultivates food from the body of the mountain. It gives four fluids (water, semen, milk, and blood) that sustain life. Mountain worship is a faith that regards mountains as sacred objects of worship.

Water is frequently presented as a supernatural being worthy of worship. Water, according to mythic accounts, is often the source of primal life. Water monsters are also the object of cultic action. They are placated or combated to stave off a repetition of the cosmic deluge. Aquatic dragons embody the fertile principles manifest in moisture. They must be slain or tamed to release their fecund powers and to prevent drought. Thus, the Chinese dragon Yin gathers together all the waters of the world and controls the rain.

The Earth is sacred in many traditions and is the object of devotion and affection. As the source of life, Mother Earth is worshipped. The agricultural cycle is coordinated with her menstrual periods, the times when she is open for conception. The Earth is frequently a partner of the sky or some other celestial fertilizing divinity.

Plants, trees, and vegetation also have their place in worship. The tree of life or the cosmic tree expresses the sacredness of the entire world.

Other kinds of vegetation also manifest sacred powers and divinities. For instance, Indian philosophy visualizes the lotus floating upon the water as a manifestation of divinity and the universe. Miraculous trees, flowers, and fruits reveal the presence of divine powers. Rites of spring frequently centre on plants, boughs, or trees that are treated as sacred. The fertility of the cosmos is symbolized by the union of male and female plants or by the blossoming of a bough from a specific species of plants. Around the world, the agricultural cycle is hedged around with religious acts directed towards the furthering of the powers of fertility manifest in various crops. In particular, the moments of sowing and reaping are marked by sacrifices. Animals have also stimulated the religious imagination in such a way as to warrant devotion. Animals, birds, fish, snakes, and even insects have all become the focus of adoration in one culture or another.

Examples of the worship of Nature could be multiplied endlessly. There is hardly any object in the natural cosmos that has not become the centre of a cult somewhere at one time or in one place or another. How this should be interpreted is a matter of extreme delicacy. In general, modern interpreters have failed to settle on a satisfactory explanation. Even the term 'Nature' carries a range of connotations that obscure the meaning of sacred objects of cult in many cultures. Each generation of scholars in the last century spawned several interpretive theories in which the worship of Nature figured as a large element in the assessment of religion in general. The effort to desacralize Nature in Western perception and to identify the perception of Nature as sacred with 'primitive' peoples played a large role in the foundation of the social sciences and self-understanding. Offering a nuanced interpretation of the worship of Nature would require a detailed deconstruction of the cultural sciences as well as a subtle appreciation of the religious terminology of each culture in question.

# 6 Complexity of Nature

Complexity is the property of a real-world system that is responsible for the inability of any one formalism to be adequate to capture all its properties. Indeed, though many conceptual models were proposed to understand the activities of Nature that manipulate its five basic elements to create and nurture the living species of the Earth, so far none of these have any convincing explanation to understand these activities.[18] Now, it is natural that the doubt 'Is it so complex to understand nature?' This kind of inquisitive doubt arises because of the curiosity to discern the true nature of the events taking place in Nature that led to the innumerable creation of animate and inanimate species we see in the world. But so far, the is no solution to this stalemate. This is because the power to analyse Nature has to come from the human mind which is also controlled by Nature. Thus, the world tries to estimate something using the power of that itself. In layman's language, this situation is something similar to a cop trying to catch himself. Thus, it appears that it is not possible to understand Nature unless Nature itself enables the mind to gain the power required for the study which aims at investigating itself.

A trivial example that can enable us to gain an insight into the complexity of Nature is the following. We all know that innocent herbivores such as deer which live on eating the vegetation, which lives in the jungle, are hunted by carnivores like lions, cheetahs, and leopards for their survival. Now, let us think of the reason for this cruel hunting of the carnivores. From their habitat, it is natural that they must hunt other animals as Nature made them survive only with meat and not with vegetation. If one thinks that being strong animals, they need to eat meat to maintain their strength, this thought is not justified, because animals that are stronger than lions, such as elephants and rhinoceros, are herbivores. Thus, it is not true that for gaining energy meat is the food.

Now, how to justify the food cycle for the carnivores and herbivores? For this involved doubt so far there is no answer, though both the scientific and spiritual worlds strive for the answer. Therefore, there is nothing other than concluding that understanding Nature is beyond the perception of the human mind. Is this conclusion final? This is the next question that would come to

DOI: 10.4324/9781003745051-6

an inquisitive mind. It is justifiably stated that this conclusion is not final and once the power embedded in the mind is stimulated it could be possible to find an answer for the complexity of Nature. The reason for this confidence is the simple fact that science keeps evolving continuously, exploring and coming out with convincing reasons for many facts that were perceived to be extremely complex.

Several examples can be seen in the world justifying the complexity of Nature that is beyond the reach of human intelligence. A typical species that can be quoted as personifying the ingenuity of Nature is the life of penguin species termed Wild Emperor penguins.[19] Wild Emperor penguins are only found in Antarctica. They breed and raise their young mostly on 'fast ice', a floating platform of the frozen ocean that is connected to the land or ice shelves. From birth, they spend their entire lives in and around the Antarctic ice, although very rarely vagrants have turned up off the coast of New Zealand. Male emperor penguins will not eat for up to four months, from the time they arrive at the colony to breed until the egg has hatched and the mother returns to feed. They lose almost half of their body weight during this time. They need to rely entirely on the reserve of body fat that they built up during the summer feast to survive the long winter. The secrecy behind the survival of this species is in the place covered with ice and the temperature can dip to a level of as low as –130°C makes the warm blood keep circulating to the foot that is in contact with ice and blood can become so cold that the colder blood will get heated up as it moves to the upper portion of the body. This kind of innovative heat transfer process makes the penguins survive in the polar region. Now, we can appreciate the power potential of Nature that is beyond the perception of the human mind.

Emperor penguins are uniquely adapted to survive these harsh conditions when temperatures can drop down to a bone-chilling –50°C and with winds of up to 200 km/hr. They have two layers of feathers, a good reserve of fat, and proportionally smaller beaks and flippers than other penguins to prevent heat loss. Emperors also have feathers on their legs, so their ankles do not get too chilly. Even their feet are adapted to the icy conditions, containing special fats that prevent them from freezing and strong claws for gripping the ice.

However, most remarkably, colonies of adults and chicks work together to huddle for warmth. Five thousand or more tightly packed adults and chicks shuffle around, so each takes a turn—but not too long—on the outside of the huddle where it is cold. This is in complete contrast to their quarrelsome and territorial neighbours, the Adélie penguins.

The continuous evolution, discoveries, and inventions in the scientific sphere are the personification of the intense complexity of Nature. The involved physical science, namely, the mind power required to resolve the fictions embedded in Nature and postulate or frame the scientific doctrine that can through light for understanding of the physical reason behind the formation or transformation of something that we encounter has to come from the brain, which is probably the most complex one that the Nature has created. Though the biological and medical worlds strive hard to understand the structure of the brain, so far there is no breakthrough in this attempt. Because of this inability to gain an insight into the structure or power embedded in the brain, the scientific world views the brain as a miraculous structure created by Nature. However, theists and atheists deem this miraculous power of the brain as the power of GOD and NATURE, respectively.

# 7 Science and Nature

Reviewing the developments in science that the world has seen may be regarded as a means to gain an insight into the power of Nature. We saw that the power required for the brain to understand anything in Nature, say the power contained in a cyclonic storm, has to come from the power that is given by Nature to the brain lobe. In other words, Nature is special in its own right that gives the energy or power required to understand its potential by itself. Science is aware that the main controller of a living species is the mind.

All the developments, say it is happiness, pain, growth, thinking power, etc., are governed by the brain. Further, the formation of the nerves and their stacking in the brain lobe is a direct indication of the power potential of the brain. Even though all sensations in a living body are controlled by the brain, the brain has no control over its activity. For instance, the mind makes a person feel the pain of an injury to any part of his body. However, if there is an injury to the brain nerves themselves, the person will not feel any pain. This insensitivity of the pain to the brain is still a black box to medical science.[6] Like this, there is an infinite quantum of events, happening, and the reason for them, taking place in Nature remains a mystery to the scientific world.

However, science continuously strives hard to understand Nature and these attempts bear fruits too. For example, when the science of thermodynamics evolved, it gave a mathematical shape to the transfer of thermal energy from one equilibrium state to another of a system.

However, though thermodynamics gave a theory to calculate the amount of thermal energy transfer required to reach one equilibrium to another of a system, it did not say anything about the state of the energy distribution in the transition state during the process of change from one equilibrium to another. The inquisitiveness to get an answer to the transients' state unanswered by the science of thermodynamics paved the path for the development of a branch of thermal science, namely 'heat transfer', that could answer the states during the transition. Soon after getting the details of the different states during the transition from one equilibrium to another, the mind urged the scientific world to learn the status of the energy distribution at every local equilibrium state in the transition process path. This curiosity resulted in

DOI: 10.4324/9781003745051-7

the advanced stage of thermal science entitled 'quantum mechanics', which found that the thermal energy at every state possesses four groups, namely (1) the translational, (2) rotational, (3) vibrational, and (4) the electronic form. Also, quantum mechanics finds a limiting temperature level up to which only the translational and rotational energy components are active and the vibrational and electronic energy will be frozen.[20] Only beyond some temperature, the vibrational mode of energy will become active and above some level of temperature at which the vibrational energy wakes up, the electronic energy will become active. Despite the capability of quantum mechanics to answer the different components of the energy and their sensitivity to the temperature level of the system, it is only able to give mathematical formulation for the calculation of the first three modes of energy and is not able to find the procedure to compute the electronic energy mode. Therefore, this limitation of quantum mechanics makes the science declare that the difference between the total energy and the sum of translational, rotational, and vibration energy components gives the electronic energy of a system.

Likewise, in almost all branches of science, there are continuous developments taking place that stand testimony to the inability of the human mind to understand Nature completely. Also, it appears that it may not be possible to understand the science fully at any stage. Does this mean that the potential of Nature keeps expanding continuously? Answering this question also is likely to remain a black box for humanity. Here comes another question, is it possible for species other than humans to understand the power of Nature fully? Some of the events such as the animals being able to understand the occurrence of some of the natural calamities like tsunamis in advance and moving to places of higher altitude and escape from the fury of such disasters make us feel that animals have a better understanding of Nature than humans.

Attempts by science to understand and control Nature are a kind of perennial wars in the world from time immemorial. Despite the continued efforts by mankind to conquer Nature, there is no convincing success so far. Even getting a glimpse of Nature's power itself has proved to be a Herculean task for the scientific world. However, though marginal, the success tasted by the science in its attempt to understand Nature inspires the scientists to continue their efforts in attaining their goal. All the inventions and discoveries seen by the world about the doctrines that govern Nature and ways to utilize the power of Nature for the welfare of the world give hope that it is possible to understand Nature to a greater extent than what is learnt so far.

Science could be able to utilize the potential of solar energy for electric power generation, in addition to the classical utilizations such as salt manufacture from seawater, and drying the natural and synthetic fabrics in the sunlight. However, the efficiency of solar panels is only from 10% to 20%. This implies that science has to go a long way in using solar power for electric power generation efficiently.

# 8 Can Science Rule Nature?

Attempting to control Nature is a continuous effort by the species from time immemorial. Despite this continued attempt, the success of this attempt is not even a few per cent. This situation raises the question 'Does it mean that it is not possible to understand nature?' For this involved doubt even the scientific world finds only a philosophical answer and not an affirmative one. Science thinks that one day it can not only understand Nature but also control it. However, the philosophical world thinks that understanding Nature is only a dream and not realistic. The reason cited for this impression by the philosophers is that the power required for the human mind to understand Nature has to come from Nature itself. This entanglement makes the situation complex.

Coming out of this complexity has proved to be a perennial problem for humanity. However, though there is no sign of even seeing a ray of light to understand the reason for this difficulty, the human race strives hard to get a glimpse of it. This persistence of the human race keeps this hope alive forever. This hope is the only lifeline for this race between humanity and Nature that poses the question of the winner.

Some success in this venture, though partial, keeps this search moving. However, philosophers think that the power embedded in Nature is limitless; hence, it is impossible to scale that mountain by the human. This kind of conflict between the scientific and philosophical worlds is also eternal in human history. Despite this conflict, science thinks that it can conquer Nature. Now the question is, is it possible for science to overcome the power of Nature that appears to be firm on retaining its supremacy? This book tries to find an answer to this inquisitive doubt.

Natural science is one of the branches of science concerned with this description, understanding, and prediction of natural phenomena, based on evidence and experimentation.[21] Natural sciences use mathematics and logic to convert information about Nature into measurements which can be explained as clear statements of the 'laws of nature'.[22] Modern natural science succeeded in more classical approaches to natural philosophy. Galileo, Kepler, Descartes, Bacon, and Newton debated the benefits of using

DOI: 10.4324/9781003745051-8

approaches that were more mathematical and more experimental in a methodical way. Still, philosophical perspectives, conjectures, and presuppositions, often overlooked, remain necessary in natural science.[23]

Systematic data collection, including discovery science, succeeded natural history, which emerged in the 16th century by describing and classifying plants, animals, minerals, and so on.[24] Today, 'natural history' suggests observational descriptions aimed at popular audiences.[25]

In natural science, impossibility assertions come to be widely accepted as overwhelmingly probable rather than considered proven to the point of being unchallengeable. The basis for this strong acceptance is a combination of extensive evidence of something not occurring, combined with an underlying theory, very successful in making predictions, whose assumptions lead logically to the conclusion that something is impossible. While an impossibility assertion in natural science can never be proved, it could be refuted by the observation of a single counterexample. Such a counterexample would require that the assumptions underlying the theory that implied the impossibility be re-examined. The theory of impossibility might be the outcome of the inability of the human mind to get a grip towards travelling on the path of its goal aiming at understanding Nature.

Now, let us try to gain an insight into the question 'whether science can conquer nature?' It may appear as an astonishingly big question to the minds without any optimism. However, for minds blessed with the capability of shedding optimistic waves, this question might even appear as an enjoyable hurdle. For instance, the first generation attempted to travel in the ocean to cross the water expanse, the only grip would have been its optimistic thoughts fuelling the fire in their mind to conquer the ocean. History reveals that in these attempts many lives were lost; many were successful too. This situation is true for all scientific ventures. If humans were scared of venturing into research and discoveries aiming at scaling the secrecies of Nature, there would not have been any devices such as flying machines, motor cars, nuclear plants, ships, submarines, missiles, and satellites.

# 9 What Is the Science Behind the Manipulation of Nature?

We see that the thermal energy contained in the coal is used for many purposes, beginning from cooking to thermal power generation. The enthalpy contained in coal is liberated as heat when it is burnt. Also, there are a few fossil fuels that are used for the generation of thermal energy. Some of the popular fuels are petrol and diesel. Another source of thermal energy is the atomic fuel. Now, let us have a closer look at these energy sources. All these are thermal energy sources. However, there is a vast difference in their energy content. What makes this difference in their energy potential is a puzzle.

Another interesting fact that can confuse us more is the difference between coal and diamond. Both are essentially carbon compounds. Coal is a carbon compound used as fuel; diamond is the hardest substance which is also a form of a carbon compound. However, there is a distinct difference between them, that is, diamond is pure carbon, and coal includes carbon but also has other elements, such as nitrogen, oxygen, hydrogen, and sulphur. It is well known that when carbon is compressed at very high pressure over a long period, the atoms contained orient in a form to become hard and shining. This hardened carbon is termed diamond. Now the question that comes to our mind is what governs this process of Nature that turns coal to diamond?

From the events we experience in Nature, such as the hatching of fish eggs during the new rains in summer, we can infer the power of Nature. The eggs, in a dried state, are embedded in the soil hatch, the fish come out soon after encountering rainwater, and the babies become active soon after hatching. Now let us look at the eggs of bird species. It is known that the eggs should be in the proper environment ensured by the birds for hatching. Otherwise, the eggs will get spoilt. In other words, for hatching, the eggs must be kept in a suitable environment for a specified period. For example, for the natural hatching of hen eggs, the hen should sit on the eggs for 21 days. The hen sits on the eggs to provide sufficient warmth to the egg for the development of the embryo into the chicks. The embryo takes about three weeks to develop into a complete chick. It is possible for science to clarify the grace of Nature which makes the embryo in fish eggs survive the heat of harsh summer and

DOI: 10.4324/9781003745051-9

come to life the moment water flows over them, and the egg of a bird gets spoilt when the environment is not congenial for its survival.

Another aspect we see in Nature that is beyond the perception of the human mind is the positioning of the life cells in the plant kingdom. Let us review a few examples. The seed of plants, such as mango, is placed inside the fruit by Nature. For some plants such as cashews, the seed is positioned outside the fruit. Now the question that comes to our mind is: can any scientific theory explain the reason behind the positioning of the seed inside and outside of the fruit? Another interesting feature observed commonly by the agriculture world is that the seeds of paddy, pulses, and so on are dried and stored for cultivation and used when the season is congenial for their cultivation. However, seeds of trees such as banyan and peepal have to be baked for their proper germination and growth. Birds eat the fruits of these plant species, and the pulp of the fruit is digested by the birds and the seeds are ejected out in their droppings. The seeds get seasoned by the heat in the stomach and become suitable for germination. This process is true for vegetables such as tomatoes. Life cycles of some species, such as potato, are different from those of other vegetables. This kind of life cycle of the plant and animal kingdoms makes it clear that what the human mind perceives is only an insignificant portion of the quantum of secrecy embedded in Nature. In this situation, what should be the mindset? Does the scientific world take a pessimistic path and hibernate or take an optimistic view and continue to forge ahead in its investigation of the secrecy of Nature? Here there is scope for both. But it will be valuable if the world resorts to the latter means. Yes, though not a smooth road to travel, it is rather a tough path to move, and it is worth continuing the journey.

Indeed, many of us know that scientists who venture into inventing anything new never bother about the hindrances that come across the path and keep forging ahead to achieve his/her goal. History shows this in the lives of successful scientists like Thomas Alva Edison and Einstein. This is true in the case of discoveries like Columbus and others who ventured into finding sea routes for the places they desired to reach.

Now comes the next big question to our mind that is more confusing than the previous questions we raised. That is, is it necessary for science to venture into assessing the power of Nature? This doubt arises because, during this probing, science interfered with Nature, leading to some serious damage to the projection extended by Nature for the comfortable life of species on Earth. For example, too much air-conditioning, vehicular emission, deforestation, etc. resulted in damage to the ozone layer covering our Mother Earth, which filters the harmful emissions. Now the world strives hard to repair this damage. This kind of event makes us wonder whether disturbing Nature is a widespread deed. There is no other way than to disturb Nature if the world wants to find new things that are perceived as comfort by humanity.

The fact that different species have different ways of survival may be perceived as the intention of Nature to keep them existing forever. If so, now we may doubt then Nature could have created and left them without ageing and dying. In this case, there would not be any infight between different species created. For the process of the birth–death cycle, the spiritual world thinks that this is the play of GOD, but the scientific world does not even give an opinion on this cycle.

# 10 Autonomy Granted to the Universe

It is widely accepted that the universe is everything that includes all of space, and all the matter and energy that space contains. It even includes time itself and, of course, it includes us. Earth and the moon are part of the universe, as are the other planets. As of today, even the size and expanse of the universe appear to be beyond the perception of science. It may convincingly be stated that the extent of the universe is larger than the mathematical unknown limit 'infinity'. In other words, it may be stated that the universe is 'space or time without end'. If this is the case, what is the basis for the scientific world's faith that the secrecy of Nature can be understood with persistent effort? Let us try to gain some understanding lying behind this faith of the scientists.

In the view of the author, the partial success gained by the scientists might be the reason behind this faith. If the inventors believed that understanding Nature is beyond human perception, as many believe, the developments leading to the invention of the many astonishing devices, beginning from electric light to satellite would not have been possible. Only with this mindset does the scientific world strive to understand the universe. These attempts keep bearing fruit, leading to the discovery of even new stars and planets. Now another important doubt might arise in our mind. That is, similar to Mother Earth, is there any other star or planet where there is life? To clarify this genuine doubt, science tries to find out the possibility of the presence of life gas, namely, oxygen and water in the members of the universe. In this search, so far only some new planets that were not known earlier have been found. However, the presence of life, gas, and water in these plants also remains an open question. Even in the moon, which is relatively closer to Earth, though there is speculation about the presence of water, there is no authentic evidence for the presence of oxygen.

National Aeronautics and Space Administration (NASA) declares that the ultimate goal of NASA's exoplanet program is to find unmistakable signs of current life on a planet beyond Earth. How soon that can happen depends on two unknowns: the prevalence of life in the galaxy and how lucky we get as we take those first, tentative, exploratory steps. This is the current status.

DOI: 10.4324/9781003745051-10

This reveals that the Earth is the only planet in the galaxy possessing life, and no extra-terrestrial life has yet been conclusively detected.

Now, coming to the power of Nature, we can elaborate on the autonomy given by Nature to the species it created with specific stipulations. Humans are perceived to be the supreme species generated by Nature with six senses, which is again believed to be the maximum number of feelings attributed to the living species. Some species have only one sense. A typical example of this case is the plant kingdom. However, the rate of growth, the size it can attain, and the life span of different species in this category vary from species to species. The mushroom grows rapidly, assumes a small size, and dies in a few days, whereas the banyan tree grows slowly, assumes a huge size, and lives for as long as 700 years. In the case of the animal world, the cat family members, lion, tiger, leopard, etc., grow fast and assume good strength but live only for about 15 years, whereas huge animals like an elephant grows gradually, becomes huge, and lives for about 100 years. This might make us think that the rate of growth and lifespan have some relation. But this reasoning is jolted by the growth and lifespan of animals such as tortoises. Some species of tortoise grow gradually, also assume a reasonably big size and live for even 700 years. These aspects demonstrate that Nature at the surface appears to be generous and gives autonomy to the lives it governs; at depth, this is not true. It may be reasonably authenticated that all the events in the life of every individual of all the species, spanning from one-sense plant kingdom to six-sense humans programmed and executed by Nature, would continue to be a mystery to the human mind. Does it mean that the puzzle of manipulation embedded in Nature will continue to be a mystery to the human world?

# 11 Doctrine and Hypothesis

The inability of the human mind to understand the mystery of the manipulation carried out by Nature in the life cycle of animate and the fury of inanimate species resulted in the divisions of science and religion, which speak about facts based on rigorous proof and intense faith, respectively. Thus, it may justifiably be stated that the theory about the power of Nature is both a scientific doctrine and a religious hypothesis. However, both these theories are perceived as useful philosophies for the peaceful existence of humanity.

As we saw, the scientific theories resulted in the design and development of many appliances that are perceived as useful for the comfortable life of the world. Similarly, the religious philosophy that revolves around the faith in the existence of some supernatural power that governs the events devised by Nature is also perceived by a large cross-section of humans. Despite the vast difference in the principles of science and religion, both are accepted by the world as useful human life. It may be stated that the main difference between science and religion is that the former is responsible for physical comfort and the latter is responsible for mental comfort.

In the scientific world, the theory and application aspects are governed by physics and mathematics. Because of this, there is no fight on any doctrine that has been accepted with proper proof and verification. However, as we know, this cannot be the case in the religious theories, as they are based on faith and superstition. Owing to this nature, there are possibilities for infight among the religions. Indeed, this is often experienced by the world. However, with the religious principle of the highest calibre, there is no scope for infighting, as this level makes the person attain a mindset that makes him free of all animosity and ego. This kind of person is the real saint. The very fact that there are real saints reveals that there are pseudo-saints also. In fact, unlike the former icons who are very few, the latter sect is large in number. This large population is responsible for all the religious chaos that the world encounters. In Nature, as per the law, all are possible and hence there is scope for all religious leaders to attain the highest level of spiritual state that would render them speechless. Despite the differences in the principles of different religious sects, the world moves without any abnormal chaos, because the

DOI: 10.4324/9781003745051-11

main philosophy of all religions is 'be kind to all'. Only this principle prevents the religious sects from indulging in violent paths. However, it should be noted that in every religious sect, some sections do not practise the actual principle of the religion they belong to and also go to the extent of hating other religious sects and going to the extreme level of clashing with another sect. Luckily, the violent act of these people is supported by the people of their religion.

# 12 The Intricacies of Nature Beyond the Reach of Science

Though numerous deeds are beyond the reach of science, some among them that are encountered by humans commonly are of great concern to the world. The one that may be considered as of primary concern to the entire humanity is of 'life' and 'death'. Science tries hard to understand what is the secrecy behind life and death. Biological science recognizes that the circulation of oxygen in the cells in the bodies of animate species is active to keep them alive. That is why oxygen is referred to as 'life gas'; the reason behind the secrecy of oxygen circulation keeping the life in a body is not known. If that is known, science could have invented an alternative means to keep a body alive forever. It is seen that over the lifetime all living species undergo an extraordinary change. But this change caused by age is explicit for some species and not so for all. For example, a person feels that his organs including the brain undergo a lot of change as he grows old. Also, these changes are noticeable. The hair which is dense and black in youth becomes thin and grey in old age. However, this kind of explicit change with age is not in the case of many animals and birds. No crow becomes white in old age.

The tight green skin of unripe mango becomes yellow and soft when it is ripened. A medium-ripe mango will be somewhat firm, and an unripe mango will be very firm to the touch. This kind of explicit change caused by ageing is seen in many species. Now the question is what is the rule behind this kind of heterogeneous changes caused by Nature? There is no answer for this kind of behaviour. For instance, the seed of a mango is inside of the fruit, whereas the seed of cashew-nut fruit is outside of the fruit. Which theory can clarify the reason for this kind of directly opposite design of Nature? Though science keeps probing into the reason for this kind of design in Nature, it would declare the reason only when it is completely understood and verified. However, this is not the case with philosophy such as astrology, palmistry, and numerology. This may be because these are not scientific, though many who practise these claim these are scientific.

We can easily justify that these do not have any scientific base. Let us look at the classification of humans under the zodiac signs: Aries, Taurus, Gemini, Cancer, Leo, Virgo, Libra, Scorpio, Sagittarius, Capricorn, Aquarius, and

DOI: 10.4324/9781003745051-12

Pisces. Is it possible to classify the fortunes of the entire world population under these 12 categories? Without any doubt, it can be stated it is not possible to do so. However, many people believe astrology is true owing to their strong belief in superstition.

The section of the human race having faith in superstition constitutes the major portion of the human population. This faith in superstition has no religious barrier. Only the shade of this faith varies from religion to religion. Astrology is the shade in the religions of Asian countries. Numerology is the shade in Western countries. Even in Buddhism, one of the major religions in China, Japan, Thailand, and Singapore, though in principle it does not believe in superstition, in practice somehow superstition crept into the minds of Buddhists.

# 13 Natural Zones Beyond the Reach of Science

The precise, fundamental reason for the existence of different human blood types remains one of the most prominent mysteries that scientific understanding has not fully explained. While science can identify, categorize, and safely transfuse blood types based on antigens, the evolutionary, "why" behind their necessity has not been entirely elucidated. In the medical field, scientists have succeeded in making blood. However, the blood that is essentially a fluid responsible for life is beyond the reach of science. Many countries attempting to make artificial blood for a long time are unable to make any noticeable progress in that direction. Though the science of flow physics has been developed at a remarkably high level in many branches, in the blood matter these advanced theories could not help scientists to make any breakthrough. Philosophically it may be stated that this is one of the zones where science retains its supremacy over human brain power. Likewise, numerous issues in Nature are beyond the reach of science so far. Even the reason for this stalemate is not clear to the scientific world. This situation reveals that Nature is above science. This may be so since science relies on human brainpower for its success and that power has to come from Nature. This simply implies that the main controller of everything in this universe is Nature.

Thus, Nature seems to retain its supremacy by not giving the required wisdom to the human mind to understand its secrecies. Despite this emphasis on Nature, science based on the human mind's power continues its journey to probe into the secrecies embedded in Nature. Only time will reveal whether science can succeed in this task.

DOI: 10.4324/9781003745051-13

# 14 Fascinating Features of Nature

Though almost all events in Nature are fascinating, the aim here is to highlight some of them we encounter in life. The first astonishing event we experience in Nature is the formation of life of all species. The species-forming cell is within all living things. It calls for the combination of male and female to give shape to this formation. This resembles the positive and negative union required for the flow of electricity. Though this rule is common for many species, there are a few exceptions to this. Another interesting aspect is that even after the amalgamation of the opposite groups, it calls for the proper environment for life to emerge. It seems that Nature is so kind to specify this environment for different species depending on their living conditions. For humans and species close to humans, these environments are slightly more sophisticated than the species that live in relatively rough environments. Also, depending on the environment's abnormal nature, species such as water-based species embryos can keep life inside until a suitable environment is encountered for this germination. The period over which this preservation can take place extends even to years in some cases. However, for humans and species close to humans, this preservation period is so short and even insignificant in some cases.

These fascinating secrecies embedded in natural events make not only the scientists but also the theists and atheists wonder about the secrecies of Nature. Though the scientific world strives to unfold these secrecies, the success so far is only partial, also in a minuscule segment such as the genesis of human, animal, and plant species. In the field of astronomy, thermal and wind power, and the presence of life in the members of the galaxies of the universe are still bewildering to scientists. As we repeatedly state, this inability of the human mind to understand Nature might be because of the simple fact that the power required for the wisdom of the mind has to come from Nature itself.

One aspect that makes us enthusiastic about continuing the effort to understand Nature more is the periodic success that the science governed by the mind waves gives to scientists. For instance, the success of Einstein occurred during his stay at the patent office, and not during his stay with Princeton

DOI: 10.4324/9781003745051-14

University as a professor. This simply indicates that the gift of Nature in the form of scientific waves to a human mind is irrespective of the position that the person holds and the environment in which he lives. Another interesting feature that highlights the power of Nature is that for animals and birds, the talent possessed by their parent comes to their offspring naturally, for example, weaving birds do whatever their parents were doing without any learning. However, these kinds of talents given to birds and animals are unilateral; unlike the skill to a person comes from their parents without practice, the new talents gained by humans can be multidimensional in the case of people blessed by Nature.

# 15 Retention of Its Supremacy

In many spheres, Nature seems to retain its supremacy. One such sphere is the production of blood. We know that our body converts the energy from the food we eat to blood. However, what is the scientific process involved in this conversion continues to remain as a black box. Indeed, scientists all over the world strive to make artificial blood, but so far no one has been successful in this attempt. Philosophically, the inability of humans to make blood may be perceived as the intention of Mother Nature to retain her supremacy. Like blood, there are numerous spheres in the life cycle where Nature retains its supremacy. Another typical example highlighting the supremacy of Nature is the life in the body of all living species. No one knows what this feature is, where it resides in a living body, or what happens to it after departing from a body. Even philosophy struggles to find an answer to this.

Philosophers feel that Nature is the supreme controller of the universe and nothing can happen without her consent. The atheists view this controller as GOD, an abstract power that can only be perceived and cannot be physically seen. Owing to the hypothetical nature of this perception, scientists simply discard this theory, and instead, they view the potential of Nature as an energy source capable of ruling both animate and inanimate species in the universe. What is it that suddenly the author says that the inanimate things are also governed by Nature? Is the doubt coming to your mind? Our doubt is genuine and the answer is also genuine. Yes, all happenings in the inanimate sphere are also governed by the power of Nature. For example, a volcanic eruption is dictated by the perturbation within the Earth. Though science could predict the eruption well in advance, there is no means to prevent it. This might be taken as evidence of the power of Nature that is beyond human perception. Likewise, carbon subjected to high pressure over some time becomes a diamond. Though science can explain the glitter associated with diamonds, there is no definite answer for the period required to convert carbon to diamonds.

In almost all spheres of Nature that science uses to study life and beyond, there are numerous species embedded in Nature that are beyond human perception. Only because of this inability of science to understand Nature

DOI: 10.4324/9781003745051-15

completely, many superstitious hypotheses in the name of GOD emerge all over the world. Indeed, this process of the emergence of new principles that hypothecates the inability of science is usually termed superstition. These activities have spread all over the world, because they serve as a means for many to live in comfort without any effort. The author does not say that all religious activities are not genuine. There are many saints and spiritual workers who do their best to serve humanity and go to the extent of stating that 'service to humanity is service to god'. What we need to keep in mind is that, like any other trade, there are genuine and bogus groups in this service also. Some of the excellent educational institutions, healthcare centres, social welfare schemes, etc. run by religious institutions all over the world are the veritable personification for the noble deed that can be deemed as Service to GOD. Seeing the service rendered by these societies might make us feel that Nature is kind this way. Only because of Nature's secrecy, do these noble sects rendering service to humanity emerge.

One of the issues that were probed into is the 'life beyond death'. For this issue, there is not even a trace of an insight. However, the theists belonging to many spiritual sects of the world think that the so-called 'soul' that is responsible for life in a body goes to a 'good' or 'bad' world depending on whether it is good or bad. They go a step further to the extent of stating that the noble among the good souls remain eternally in the good world termed 'heaven' and the bad souls suffer for a period in the demon world termed 'hell', and then take rebirth.

Numerous examples can be given of the supremacy of Nature, which is beyond the understanding of science. A typical example of this is life in the underwater world. It is known to science that most of the plans require soil that is salt-free for their growth. Also, the water used for irrigation should be free of salt and hence seawater is not suitable for farming. However, numerous plants grow on the seafloor bed and feed many species living in the sea. Though there were attempts by science to understand this peculiar physics, there is no answer to this wonder.

# 16 Paradise and Perdition

Paradise is perceived as the world of happiness, where noble souls will lead happiness without ageing, illness, and other pains. Also, it is viewed as the world of brightness. The opposite of paradise is hell. Hell is the world of darkness and the souls of sinners will go into this world and will be punished for their bad deeds. This perception of a good and bad world is common to almost all the sects of the religion practised in the world. Here it should be emphasized that not all religions believe in the rebirth theory.

Eastern religions like Hinduism, Jainism, and Buddhism have differed in their faith in rebirth. Further, Islam as well as the most dominant religion of the world, Christianity, having its origin in the West, has largely denied reincarnation, though some subsects still show interest in it. We may state that there is no evidence to ascertain the theory of rebirth. Similarly, there is no evidence to set this aside too. In this situation, it is wise to ignore this debate and move ahead with our discussion on the power and autonomy of Nature, which is perceived as the main source of the action of the universe by science.

In the view of the author, the waves emitted by a mind focusing on science primarily contain the scientific wave spectrum and the spiritual content is minimal. This is because no mind can shed waves that are completely free of scientific or spiritual content. Therefore, it is bound to be a blend of both. However, the domination of the scientific or spiritual content varies from one mind to another. The result is that the mind-shedding waves that are rich in scientific content result in invention or discovery and the mind-shedding waves with rich spiritual content result in meditation, prayer, etc., which focuses on spiritual enlightenment. Based on this view, the world can be classified into two. The Western world appears to be rich in the scientific sphere and the Eastern world is rich in the spiritual sphere. This division is not rigid. It is common to see that there are contributions to science in the Eastern world and contributions to the spiritual sphere in the Western world.

DOI: 10.4324/9781003745051-16

# 17 Making Use of Natural Wealth

So far, our discussion has lingered around the power and autonomy of Nature. Now, let us look at the feasibility of making use of Nature for the welfare of the world. While analysing the aspect of utility, it is essential to keep in mind that for every action there is a reaction. Our action here is to probe into the process to make use of the power of Nature. During this utilization, the reaction could be harmful. For example, in nuclear applications such as atomic power generation, though the energy liberated as thermal energy is enormous, the fuel burnt for this emits harmful radiation that can prove to be disastrous if not prevented from getting into the environment.

Nature, especially scenes with high fascination value, encourages the replenishing of mental energy by allowing the mind to function somewhat effortlessly.

Connecting with Nature can have many positive effects on mental and physical health, including the following:

1. *Improved mood:* Spending time in Nature can generate positive emotions like joy, calmness, and creativity.
2. *Reduced stress and anxiety:* People who spend time in Nature tend to have lower levels of stress, anxiety, and depressive symptoms.
3. *Better concentration:* Nature can help with concentration.
4. *Improved physical health:* Nature can reduce blood pressure, heart rate, muscle tension, and the production of stress hormones.
5. *Increased confidence and self-esteem:* Spending time in Nature can help improve confidence and self-esteem.
6. *Increased activity:* Nature can help people be more active.
7. *Reduced loneliness:* Nature can help reduce loneliness.
8. *Increased pro-environmental behaviours:* People who are more connected to Nature are more likely to engage in pro-environmental behaviours like recycling and buying seasonal food.

Even a simple plant in a room can have a significant impact on stress and anxiety. More remote and biodiverse spaces may be particularly helpful, but even urban parks and trees can lead to positive outcomes.

DOI: 10.4324/9781003745051-17

# 18 Synchronizing with Nature

Despite the desperate efforts to lead a healthy life, people all over the world suffer from numerous ailments. Though the scientific world would have a different view than the view expressed here, it is worth reviewing this reason cited for the suffering of the world. If we observe the life of species other than humans, the view expressed here as the reason for the suffering of humans can be understood.

In the animal kingdom, it is common that most of the species lead a healthy life throughout their life span. The main reason for this is their habits. The animals—whether they are carnivores, herbivores, or omnivores—eat only the food available in their living environment. For example, herbivores like elephants survive on the plants growing in their environment. Similarly, the animals such as giant pandas live mostly on bamboo leaves. Neither the African elephant desires to taste bamboo nor the panda longs for grass that grows in Africa. Likewise, animals living in water survive mainly on fish that live in water and land animals, such as lions, live on meat of deer, buffalo, and other animals found in their living place. Neither of these animals long for the food in another environment other than theirs. Because of this synchronization with the natural environment, the animals do not suffer any unwanted ailments. What is meant by the adjective 'unwanted' implies ailment caused by food other than that available outside the habitat environment.

In the case of mankind, most of the health issues are associated with eating something that is not grown in its place. For instance, rice is a natural gift to most Asian countries and wheat is the food decided by Nature to Europeans. However, owing to globalization, there is a lot of migration from one place to another that has different weather and lifestyle. Also, there is a kind of fascination with eating a variety of food without even identifying their origin. This habit is the genesis of most of the diseases that the world encounters nowadays. Another main cause for these ailments faced by the human race is the pharmaceutical agencies. It is common to see advertisements by pharmaceutical companies that almost threaten the world stating that without those drugs life would become impossible.

Now, it is common to ask, why should the world listen to these if they are not right? The answer to this simple question is, not all people are

DOI: 10.4324/9781003745051-18

strong-willed. People with strong minds do not yield to these advertisements and simply ignore them. These are blessed with wisdom and their wisdom made them lead to an unperturbed life. To highlight this, there is a simple story that many of us might have heard of. That is the following:

*Story*: It is common in GOD's court that every year the demon should kill a specified portion of the population to keep it under control. According to this, one day the death controller came to GOD seeking permission to kill the percentage of the population specified by the Lord for that year.

*Demon*: My Lord please specify the quota for this year for my execution.

*GOD*: This year you can kill 10 per cent of the population.

The demon went out with this order and began his task. But at the end of execution, the number of people killed in that year amounted to 90 per cent of the population. GOD was annoyed about this inappropriate proportion and summoned the Demon. The Demon came to the court and bowed at GOD.

*GOD:* Asked with fury, why did you kill 90 per cent, whereas I permitted you to kill only 10 per cent?

*Demon:* My Lord, as you are aware I am sincere in my duty. As per your direction, I spread the epidemic to kill 10 per cent of the population and completed only that. But the rest of the 80 per cent population died this year not because of my actions but due to their own fears.

This story should be taken as the outcome of the social mindset. In fact, the majority of human society falls into this category. Strong-willed people are always a minority.

Among the rationalists, it is perceived that the pharmaceutical world spreads news about new viruses almost every year aiming at causing panic to the world. Soon after succeeding in this attempt, it advertises that this virus can be controlled with this new drug developed. The next stage is to work on a tie-up with the governments of different countries in the world to make them come to an understanding, of course with a hidden agenda so that the power centre of these nations will make people take this preventive medication. This might look natural at the surface level, but the undercurrent in this episode is something that cannot be quantified.

Naturalists make observations of the relationships between organisms and their environments, as well as how those relationships change over time. One of the most well-known examples of a naturalist is Charles Darwin. These observations reveal that synchronizing with Nature is the key to a healthy life. This feature is true for all species. This kind of study is termed 'ecology'. Ecologists study how organisms and their surroundings interact.

It is preached by both scientists and saints that tuning with the natural cycle is the key to liberating the power potential of the mind and using it for productive purposes. Let us have a closer look at the word 'Tuning'. Tuning simply means adjusting to the time cycle of Nature. In tropical zones, the difference between night and day duration is appreciable. However, at higher latitudes, the day and night duration can even be to the extent of 24 and 0 hours during the summer solstice and 0 and 24 hours during the winter solstice. Therefore, it becomes imperative on the part of people in places with large variations in day and night to tune their minds to this vast variation in day and night, to perform efficiently in their activities. However, this need is not observed for animals, birds, and other species living in such an environment. For example, polar bear hibernates during winter, because there is no need for them to be active in its duty like humans.

Another need for humans during this kind of drastic change in the period is to keep their body at the proper temperature by eating suitable food. In winter, people need to eat food that will generate a lot of calories to keep their body warm and during summer the same people need to eat food that will cool off their body.

This kind of synchronization with Nature is essential for the survival of all living species.

# 19 Beyond Science

Some aspects are beyond the understanding of science. In botany, it is a surprise that there are two distinctly different categories. In one group, their seed is located inside the fruits, whereas in some specific groups like cashews, the seed is outside the fruit. Similarly, in most of the species, the yield begins with buds leading to the formation of results. For example, in vegetables like eggplant, the flower comes first and the vegetable formation takes place later, whereas in vegetables such as pumpkin the baby vegetable formation and the flowering take place simultaneously. Science thinks that the reason for this kind of feature is secrecy which is unique to Nature, probably to explicitly exhibit its supremacy.

In humans also, there are numerous aspects governed by Nature, which are beyond scientific understanding. For instance, in the case of the animal kingdom, plant kingdom, etc., the nature of a particular species is almost alike. However, in humans, the children born to the same parents have different characteristics, not only in their mental stature but also in their physical features. Science to some extent assigns the mind state for the father and mother, at the instant of semen's entry to the womb. Again, this is not proved with mathematical relations or theory of physics.

Likewise, there are numerous spheres in the universe with secrecies that are beyond the reach of science. Indeed, the science itself is of the view that it has to go a very long way to understand Nature completely, though it does not declare that this venture is beyond its scope. However, the philosophical and spiritual spheres think that it is not possible to understand Nature fully. This view of these spheres may be because, despite continuous efforts by scientists to gain insight into natural calamities such as seismic activity, etc., there is only partial success in these attempts.

It is common in the minds of most that I alone experience these many problems and the rest of the world happily enjoys life. The inner trace in this kind of thought is that this feeling is in the mind of almost everyone, though there are a few exceptions to this. Now the question is what is the reason for this kind of mind state? Anthropology tries its best to find an answer to this. But there is no answer to this question.

DOI: 10.4324/9781003745051-19

Lord Buddha said, 'desire is the cause of misery to the human'. This is believed to be the outcome of the wisdom attained by Buddha. But though in principle this might be acceptable to all the sects of the world, relinquishing desire is not easy for a common man. Also, even among other species, desire seems to be the source energizing their actions. For instance, animals are motivated to hunt by their desire to eat. If they overcome this urge and rest in peace without eating, there is no need for them to hunt. The pathos is that the animals do not hunt if they are not hungry, but humans hunt for wealth even when there is no need. This is because of greediness. Why does not Nature make the human mind think like animals? This question has been raised by nobles all over the history. However, there is no wisdom gained from this secrecy hidden in Nature, even by the great souls that history has seen.

Urging to understand Nature is a continuous process. Though these attempts by humans strive hard to gain an insight into the reason for many events taking place in Nature to protect from this fury and utilize the power of Nature for the comfort of the world, the situation so far is something like, 'for every yard climbed there is a slip of three yards'.

# 20 Nature and Human Life

The combined effect of genetics and environment significantly influences human development. The interactive role played by these two components is crucial for human life.[26] While genetics provide the foundation for physical and psychological traits, environmental factors like parenting, culture, and experiences can shape how these traits are expressed and develop. Human nature encompasses a wide array of traits, including social tendencies, emotional capacity, cognitive abilities, and the desire to learn and grow. These qualities, which are often universal, influence how individuals behave, interact, and perceive the world.

Genetics, the biological makeup inherited from parents, sets the stage for many aspects of human development, including physical traits (like height, eye colour), temperament, and susceptibility to certain diseases. Studies carried out on twins reveal that even identical twins, despite sharing the same genes, can develop slightly different personalities or have different levels of certain cognitive abilities, suggesting that genes are not the sole determinants of development.

The environment encompasses a wide range of factors, including family, culture, education, social interactions, and even the physical surroundings. A child's language development is heavily influenced by the language spoken in their home and their interactions with other people, while exposure to enriching educational experiences can significantly impact cognitive development.

The environmental factors can influence gene expression, meaning they can affect which genes are turned on or off, which can have long-term consequences for development. A child's temperament, for instance, can be influenced by their genetic predisposition, but their experiences with parenting style and social interactions can also play a role in shaping their temperament.

Parenting style has a significant effect on the behaviour of a child. The social skills of a child can be nurtured through positive interactions with peers and adults, and their ability to form healthy relationships can be influenced by their early experiences.

The natural environment has a dominant role on the growth and shaping of the body and mind of a person. Indeed, the environment significantly

DOI: 10.4324/9781003745051-20

affects the human life, influencing our health, well-being, and survival. Factors like the quality of air and water, climate, and the presence of pollutants can have direct health consequences, while environmental degradation can disrupt ecosystems and impact food security. Also, the environment shapes our physical and mental health, influencing our behaviour.

A natural environment for mental health includes green spaces like parks and forests, as well as blue spaces like rivers and beaches, which provide a calming and restorative experience. These environments offer benefits like stress reduction, improved mood, and enhanced cognitive function. Connecting with Nature, even through activities like gardening or watching Nature documentaries, can also positively impact mental well-being. Environments such as parks, forests, and gardens can provide mental health benefits. Rivers, lakes, beaches, and wetlands can offer calming and restorative experiences. Growing plants or flowers in our house can be a fulfilling and rewarding experience that can boost mental health.

Observing birds, animals, or other wildlife can be a calming and engaging activity. Formal treatment involving activities in Nature, like ecotherapy, has shown promising results in relief from mental as well as physical stress.

# 21 Mysteries in Nature

This chapter emphasizes that many everyday natural phenomena—like life, death, and weather extremes—remain mysterious despite scientific advances. It underscores the enduring wonder and complexity of Nature, much of which is only partially understood or still unexplained.

Despite continued efforts by the scientific world, there are numerous events in Nature that are beyond the perception of science. Some of the well-known events that are commonly encountered in day-to-day life that continue to be miraculous are life and death, lightning and thunder, fruits acquiring sweetness on ripening, and the pleasure derived on watching some pleasing events and sun are a few events which were either only partly understood or not understood at all. A simple example to highlight the mystery of Nature is the extreme weathers that prevail in the Universe.

For instance, deserts experience extreme temperatures, both hot and cold. Some deserts can reach scorching temperatures during the day (up to 50°C or 122°F), while nights can be extremely cold, sometimes near freezing. Some of the well-known deserts are the ***Sahara Desert:*** The largest hot desert in the world, known for its extreme temperatures and sand dunes; the ***Gobi Desert:*** A large, cold desert in Asia, experiencing harsh winters and hot summers; and the ***Atacama Desert:*** A desert in South America, known for its extremely low rainfall. Now the question is what is responsible for the life of vegetation, animals, and humans in the desert. Due to the lack of rainfall, vegetation in deserts is sparse. However, there are some desert-adapted plants, like cacti and shrubs, which can survive in these harsh conditions. Animals living in deserts have various adaptations to survive the heat and lack of water. Some burrow underground to escape the daytime heat, while others obtain moisture from their food. Humans who live in deserts have developed ways to cope with the harsh conditions, including wearing loose-fitting clothing to protect from the sun and storing food and water. There is no definite explanation for the adoption of life in the extremely harsh environment of the desert. For example, desert scorpions have adaptations that help them survive in dry conditions, such as the ability to survive without food for

DOI: 10.4324/9781003745051-21

extended periods and specialized leg bristles for walking on sand and building shelters that provide shade.

There are numerous mysteries hidden in Nature that are beyond the perception of humans. Even the intellectuals with extremely high intellectual ability compared to an average human mind such as Sir Isaac Newton, who derived all his inspiration from the Nature for his inventions, understood only an insignificant portion of the secrecies hidden in Nature.[27]

Nature is brimming with hidden secrets, often unnoticed by us, revealing intricate connections and wonders in the natural world. These secrets range from microscopic details to large-scale ecosystems, highlighting the delicate balance and surprising interactions between species and their environment. The world of bacteria and other micro-organisms is teeming with life, playing crucial roles in various processes, from decomposition to nutrient cycling. Animals and plants have evolved incredible adaptations, including camouflage, mimicry, and specialized behaviours, to thrive in their environments. Hidden beneath the surface, geological processes like tectonic plate movement and erosion shape the Earth's landscapes and create unique habitats. Researchers continue to uncover new information about natural systems through various scientific disciplines, including ecology, genetics, and microbiology. Discovering Nature's hidden secrets fosters a sense of awe and wonder, encouraging us to connect with and protect our planet.

Some of the most interesting, unexplained phenomena and occurrences are the following. The significance of the golden ratio to natural floral systems (approximated as 1: 1.618) is a mathematical concept that appears frequently in Nature and is often used in floral design to create visually appealing and harmonious arrangements, the design, and the reason behind the same.

Abiogenesis is the scientific study of how life originated from non-living matter. Why is Planet Earth called the 'Pale Blue Dot'? How did it manage to accumulate so much water on its surface over millennia? The exploration of consciousness has long fascinated philosophers and scientists, driving the quest to understand how the brain produces subjective experiences. This pursuit encompasses the study of qualia and the neural correlates of awareness, posing significant challenges in science. Gaining insight into consciousness has profound implications for our self-understanding and our place in the universe. The truth behind evolution and the true, accurate branching of the *Homo sapiens*—the biological classification of modern humans is a subject of continued research for scientists.

## 21.1 Reason for the Extreme Conditions

If one thinks of the reason or the need for creating or establishing the extreme states such as scorching heat and freezing cold on the stars and planets of the universe, the possible conclusion to this inquisitee doubt is 'it is a miracle' or the answer to this question is beyond human perception.

The existence of extreme environments in the universe, from superheated black holes to frigid space, can be philosophically understood through the lens of natural selection and the broader concept of possibility. The universe is not a curated 'life-friendly' place; rather, it is a vast, chaotic landscape where a wide range of conditions exist, some of which may be conducive to life as we know it, and others may not, or even pave the way for different forms of life entirely. The universe, from a philosophical perspective, is not just what exists, but what is possible. Extreme environments highlight the vast range of potential conditions that could support life, even in ways we cannot yet imagine. For instance, the expanse of the universe itself is a mystery. Only some surface-level information is available to the world despite the continued effort by the scientists over a long period. The very fact that space scientists come out with new stars and planets reveals that a complete understanding of the universe is beyond the reach of science. Likewise, continuous exploration of animate and inanimate things by science shows that there is a lot more to understand by us. These aspects authenticate the mystery of Nature.

Now the question is who is the controller of Nature? As envisaged by the religions, is the 'Super Power' or 'GOD' the controller of the Nature power? If so, is it possible for saints to gain knowledge of the complete power potential of Nature? An important aspect to be borne in mind is the concept behind religion and science. Religion or spiritualism is a philosophical principle and hence can only be perceived without any physical verification, whereas science is based on the concepts that can be verified physically. Because of their nature, there is a continuous conflict between science and religion.

# 22 Unsolved Wonders in Nature Mysteries

Nature presents many mysteries, ranging from fundamental scientific enigmas to peculiar regional phenomena. These mysteries include *the nature of consciousness, the existence and composition of dark matter and energy, the origins of life, the Fermi Paradox, the Tunguska event, ball lightning, and the Mpemba effect.*[28] The nature of consciousness, the subjective experience of awareness and thought, remains one of the most profound mysteries in science. These invisible substances make up the vast majority of the universe's mass and energy, but their exact composition and nature are unknown. This phenomenon, where particles become linked and instantaneously affect each other regardless of distance, raises profound questions about the nature of reality.

One of the evergreen subjects of scientific research is the 'hidden treasures of mind'. This refers to the untapped potential, knowledge, and wisdom that individuals possess but may not realize or fully utilize. It is about recognizing the valuable resources within ourselves, including hidden talents, strengths, and inner resources that can be discovered and developed to enhance one's life. This idea transcends various traditions and cultures, symbolizing the potential for esoteric knowledge, untapped resources, and spiritual insights.[6]

In various spiritual and religious traditions, including Buddhism and Jainism, hidden treasure symbolizes esoteric knowledge, spiritual insights, and untapped potential. It can also represent the wealth of inner resources and talents that individuals possess but may not be aware of. The idea of hidden treasures in the mind encourages self-reflection, personal growth, and the pursuit of knowledge and understanding. It suggests that individuals can unlock their full potential by exploring their minds and developing their hidden strengths and talents. This process of self-discovery often involves introspection, meditation, learning new skills, and engaging in activities that spark curiosity and creativity. It can also involve seeking guidance from mentors or exploring different fields of knowledge to uncover hidden interests and passions.

Beyond the practical application, hidden treasures in the mind can also be interpreted in a more transcendental way, representing the potential for

DOI: 10.4324/9781003745051-22

spiritual enlightenment, inner peace, and a deeper understanding of oneself and the world.

The concept of hidden treasures has been explored in various historical and literary contexts, such as tales of buried riches, hidden libraries, and lost artefacts. These stories often serve as metaphors for the hidden potential and wisdom that can be discovered within oneself.

## 22.1 Is It Possible to Control Nature?

As we reflect on our relationship with the natural world, it becomes clear that ***we do not hold ultimate control over the environment.*** Our power lies in our capacity to understand, respect, and work in harmony with Nature's intricate systems. In fact, humans controlling (manipulating) Nature and ruling over (manipulating) plants and animals has brought us to the high place we are today.

Humans are part of Nature, so in that sense, Nature always does its thing.

Human behaviour certainly affects the rest of Nature—we have caused plenty of large-scale extinctions, deforestations, not to mention climate change itself.

In the strong sense, we certainly do not control these things—they were not exactly intended, and it is quite ambiguous whether we could have done differently, given our natures, history, etc. This does not mean that we should always just blunder along!

Human dominion over Nature is a fallacy. Although we would like to think we have mastered Nature, this cannot be because we are part of Nature; we came straight out of Nature. Humans are inimical to Nature because we are the most confused species there is. Nature is our biotope or habitat, but we are in the process of bleeding it to death. Out of all the species, we are the only one that cannot 100% adapt to Nature. So, we turn against it.

However, nature is actually more powerful than humans. Every now and then, it reminds us who is the boss, thanks to volcanoes, cyclones, earthquakes, tornadoes, tsunamis, floods, lightning strikes, and wild fires.

Humans will never outlive Nature because we cannot exist in a vacuum. So, long after our species is gone (like it was the case for every single species who lived on the surface of the Earth before us), Nature will be there.

Sometimes we try and are successful for a short period of time, like damming rivers to make electricity. Eventually every dam will fail, however. Nature moves the tectonic plates on the surface of the Earth. Nature produces hurricanes, the tides, volcanoes, earthquakes, and tsunamis. We are helpless before the forces of Nature.

## 22.2 Human Mind Action

The behaviour of mind in a human body is one among the surprising mysteries of Nature. For example, the inability to recall events before a certain age, typically around 3–4 years, is known as childhood amnesia or infantile amnesia. Several factors contribute to this phenomenon, including the maturation of brain structures involved in memory, the encoding of memories, and the development of a self-concept. The hippocampus, responsible for forming new memories, and the amygdala, involved in emotional responses, continue to develop significantly during early childhood. These brain regions are crucial for encoding and retrieving memories.

As the brain matures, it undergoes synaptic pruning, where less-used connections between neurons are eliminated, potentially leading to the loss of some early memories.

Early memories are often pre-verbal, meaning they are not tied to language. This makes it difficult for us to access and retrieve these memories later, as we tend to rely on language and symbolic representations when recalling events.

Memories before age 4 are encoded differently, possibly in a more sensory and emotional way. This difference in encoding might contribute to the difficulty in later recall.

The development of a self-concept and a sense of autobiographical identity is crucial for forming and retrieving long-lasting memories.

Memories from before age 4 are often based on stories and pictures, making them susceptible to being altered or even fabricated.

The age of first memory varies significantly between individuals, and it is not uncommon for people to misdate their earliest memories.

Memory changes with age. As people grow older, changes occur in all parts of the body, including the brain. Our ability to remember new information peaks in our 20s, and then starts to decline noticeably from our 50s or 60s. Because the hippocampus is one brain region that continues producing new neurons into adulthood, it plays an important role in memory and learning.

The 'sharpest' brain is not defined by a single age, as different cognitive abilities peak at different times. While some skills like raw processing speed may peak in the late teens, others like emotional understanding continue to develop into the 40s and 50s. Overall brain processing and detailed memory generally peak around age 18, but the ability to learn new information peaks around age 50 years.

A child's brain develops up to 90% by age 5! At this age, their brain is almost fully formed, making the first five years a crucial opportunity for developing connections that power the brain and impact lifelong learning and success.

AI, or artificial intelligence, aims to create machines that can think and learn like humans, encompassing tasks like perception, speech recognition, decision-making, and language translation. The field also includes efforts to simulate the human brain, often referred to as brain AI simulation, with the goal of creating computational models that mimic our intricate organ's structure and function.

## 22.3 AI and Human Mind

AI is a broad field that utilizes various technologies, including machine learning and deep learning, to enable computers to perform tasks that typically require human intelligence. AI systems are designed to mimic human cognitive abilities such as learning, comprehension, problem-solving, and decision-making. It is used in diverse applications, including visual perception, speech recognition, natural language processing, and more.

The quest to simulate the human brain involves creating computational models that mirror the structure and function of the brain. Human brain AI simulation has real-world implications, potentially leading to advancements in areas like disease diagnosis, treatment, and cognitive enhancement.[29]

Human and AI differ in significant ways, however. They are not synonymous or fungible. Even given the still intense internal contestation over what defines human intelligence and what defines AI, the differences between the two are clear. There are two primary differences between AI and human intelligence. They are:

- AI requires millions of samples to learn at a level exceeding average human intelligence. This makes human intelligence more efficient than AI.
- The capability of human mind to form ideas, mental sensations, and concepts of a phenomena without the physical presence of anything is unique, but this is not possible for AI.

## 22.4 The Physical Constitution of Humans and Animals

The physical constitution of humans and animals encompasses the unique combination of physical and physiological characteristics that define an individual's body and its functions, influencing their health, behaviours, and interactions with the environment. This constitution, determined by genetics, environment, and lifestyle, plays a crucial role in both individual and species-level survival and adaptation.

The main features of physical constitution encompass the genetics. Inherited traits from parents, including physical appearance, metabolism, and susceptibility to certain diseases, significantly influence an individual's constitution. Factors like climate, food availability, and exposure to toxins can alter an individual's physiology and physical development over time, and habits such as diet, exercise, and substance use can have a lasting impact on health and physical constitution. The ways how the body processes energy, responds to stimuli, and maintains homeostasis are all part of an individual's constitutional makeup. An individual's ability to withstand stress and recover from illness is also linked to their constitution.

Different ethnic groups of humans exhibit varying physical characteristics due to their evolutionary history and adaptation to specific environments.

Animal species have evolved distinct physical traits, such as fur thickness in cold climates or camouflage in predator–prey interactions, which enhance their survival.

Physical constitution is a key factor in how species evolve and adapt to changing environments. Certain diseases are more prevalent in individuals with specific constitutions, highlighting the importance of genetic and environmental factors.

Humans and animals share a fundamental physical composition based on basic elements and molecules, though specific details vary. Both are primarily composed of water, protein, and fat, with trace amounts of minerals. Human and animal bodies are also made up of six main elements: oxygen, carbon, hydrogen, nitrogen, calcium, and phosphorus.

Oxygen, carbon, hydrogen, nitrogen, calcium, and phosphorus make up the vast majority of body mass in both humans and animals. Smaller amounts of other elements, such as potassium, sulphur, sodium, chlorine, and magnesium, are also essential for various bodily functions. Water content varies in different tissues. For example, blood plasma contains a high percentage of water (90–92%), while bones contain a lower percentage (45%).

Animals and humans have similar organs (like the heart, lungs, brain) and organ systems (like the respiratory, cardiovascular, and nervous systems) that perform the same functions. One key difference between humans and animals is the development of muscles and mobility. Animals are known for their diverse range of movement, which is crucial for foraging, hunting, and reproduction. Humans have developed the capacity for complex thought, logic, and reasoning, which sets them apart from most animals.

Humans and animals share fundamental biological similarities, but key differences exist in body constitution, including locomotion, digestive systems, and brain complexity. Humans are bipedal; walk upright on two legs, have a single-chambered stomach (monogastric), and possess a larger, more complex brain, enabling reasoning and language. Most animals are quadrupedal; quadrupedal walk on four legs, although some animals have different locomotion methods (e.g. reptiles crawl, aquatic animals swim), have specialized digestive systems (e.g. ruminant systems), and are primarily driven by instincts.

The brain of humans is larger and more complex, enabling abstract thought, language, and reasoning. But the brain size and complexity of animals vary, with most relying primarily on instincts. Humans have true language for their communication, but animals communicate through some methods that lack the complexity and expressions of human language.

Humans can manipulate and alter their environment to suit their needs. But animals adapt to their environment for this survival; otherwise, they cannot survive.

# 23 The Outer Space

The outer space is part of the universe. The universe, consisting of stars, planets, and galaxies, is considered the natural world.

Space provides the context in which physical laws operate and governs the behaviour of matter and energy. In a broader sense, Nature includes everything in the universe, including both physical components (like matter and energy) and the empty regions of space that separate them.[30]

The discovery of extremely ancient galaxies, the potential for life beyond Earth, or the ability to overcome seemingly insurmountable technological challenges is an indication of the limitation of the present science forbidding the complete understanding of Nature. However, the hope that one day the hidden secrecies of Nature can be explored by science is the basic energy that keeps science forging ahead in its attempt to understand Nature with the aim of using its power for human comfort. The ability for humans to travel to and live in space, despite the harsh conditions, overcoming challenges like the effects of microgravity on the human body, is one such confidence-inducing factor that keeps the search alive.

After gaining an insight into the activities of the natural inhabitants of the Earth, science began its search on the outer space that exists beyond the Earth's atmosphere. In this search the first task encountered is to find an answer for the doubt, Whether the outer space and its constituents obey the laws of Nature? It is envisaged that all parts of the universe, including outer space and its constituents, are believed to be governed by the same fundamental laws of Nature. This includes gravity, electromagnetism, and the laws of atomic and nuclear physics. These laws are thought to apply universally, meaning they operate consistently across all locations and times in the universe. The fundamental laws of Nature, such as gravity, are believed to apply universally, meaning they operate consistently across all locations and times in the universe.[31]

DOI: 10.4324/9781003745051-23

# 24 The Science of the Universe

The science of the universe, termed cosmology, encompasses the study of its origin, evolution, structure, and future. It draws from various scientific disciplines like astrophysics, physics (nuclear, particle, and plasma), relativity, and quantum mechanics. Key areas of study include the Big Bang Theory, the formation of galaxies and stars, dark matter and dark energy, and the ultimate fate of the universe.

The Big Bang Theory[32] is the prevailing cosmological model, suggesting the universe originated from an extremely hot, dense state about 13.8 billion years ago. This singularity rapidly expanded, cooled, and formed the universe we observe today. The theory is supported by evidence like the cosmic microwave background radiation and the abundance of light elements.

The universe is vast and structured, with galaxies, stars, and planets organized into clusters and superclusters. Dark matter and dark energy play a significant role in the universe's structure and evolution, though their nature is still not fully understood. Dark matter is thought to make up the majority of matter in the universe and provides the gravitational scaffolding for galaxies. Dark energy is believed to be responsible for the accelerating expansion of the universe.

The universe is constantly evolving, with the ongoing formation of stars and galaxies. The first stars formed around 180 million years after the Big Bang, with galaxies appearing around 300 million years after. The universe's expansion is accelerating, a phenomenon attributed to dark energy.

The universe made carbon, calcium, and oxygen necessary to construct something as interesting and complicated. Today, many billions of years later, the universe has evolved into a more hospitable place for life. Science is not merely a body of knowledge, but a method by which we attempt to understand Nature and how it behaves. This method begins with many observations over a period of time. From the trends found through observations, scientists can model the particular phenomena they want to understand. Such models are always approximations of Nature, subject to further testing.

Over centuries scientists have extracted various scientific laws from countless observations, hypotheses, and experiments. These scientific laws are, in a sense, the 'rules' of the game that Nature plays. In astronomy, we deal with distances on a scale you may never have thought about before, with numbers

DOI: 10.4324/9781003745051-24

larger than any you may have encountered. We adopt two approaches that make dealing with astronomical numbers a little bit easier. First, we use a system for writing large and small numbers called scientific notation (or sometimes powers-of-ten notation). This system is very appealing because it eliminates the many zeros.

This sets a limit on how quickly we can learn about events in the universe. If a star is 100 light-years away, the light we see from it tonight left that star 100 years ago and is just now arriving in our neighbourhood. The soonest we can learn about any changes in that star is 100 years after the fact. For a star that is 500 light-years away, the light we detect tonight left 500 years ago and is carrying 500-year-old news.

Learning astronomy is a little like learning a new language: at first, it seems there are so many new expressions that we will never master them all, but with practice, we soon develop facility with them. At this point, we may also feel a bit small and insignificant, dwarfed by the cosmic scales of distance and time. However, there is another way to look at what we have learnt from our first glimpses of the cosmos.

## 24.1 The Universe by the Religions

Religious cosmologies offer diverse perspectives on the universe, often differing from scientific cosmology by including spiritual and moral dimensions. While science focuses on empirical observation and theories, religions often explain the universe's origins, purpose, and human place within it through narratives, traditions, and metaphysical systems.

***Judaism, Christianity, and Islam*** believe in a divine creation of the universe by GOD, often with a specific beginning and order of creation. The *Book of Genesis* in the Bible, shared by Judaism and Christianity, details GOD's creation of the heavens and the Earth. Hinduism views the universe as cyclical, undergoing endless cycles of creation, preservation, and destruction, with each cycle lasting for vast cosmic ages. Buddhism focuses on the interconnectedness of all things and the impermanence of existence, rather than a singular creation event. Jainism believes in eternal substances that make up the universe, existing in cycles of creation and destruction.

Many religions posit a divine creator or a higher power that brought the universe into being and continues to govern it.

Religions often offer explanations for why the universe exists and what role humans play within it, providing a framework for moral behaviour and spiritual growth.

Concepts of the afterlife vary widely, from heaven and hell in Abrahamic religions to cycles of rebirth in Hinduism and Buddhism.

Religions often include ethical guidelines and moral codes that are believed to be aligned with the natural order of the universe.

Science relies on empirical observation, experimentation, and testable theories. Religious cosmologies often draw on revelation, tradition, and faith.

Scientific cosmology focuses primarily on the physical structure and evolution of the universe. Religious cosmology often encompasses broader questions about meaning, purpose, and human destiny.

#### *24.1.1 Religious Cosmologies*

Hinduism emphasizes cyclical time, the interconnectedness of all things, and the concept of Brahman, the ultimate reality.

Buddhism focuses on the impermanence of all things, the importance of overcoming suffering, and the path to enlightenment.

Jainism believes in the existence of eternal substances and the concept of karma, which determines one's future rebirths.

Taoism emphasizes the Tao, the fundamental principle of the universe, and the importance of living in harmony with Nature.

### 24.2 Origin of the Universe

The most widely accepted theory for the origin of the universe is the Big Bang theory. This theory posits that the universe began as an extremely hot, dense point that rapidly expanded approximately 13.8 billion years ago, creating space, time, matter, and energy.

The universe is thought to have started from a singularity, a point of infinite density and temperature. This singularity underwent a period of rapid expansion, known as the Big Bang.

As the universe expanded, it also cooled, and this cooling allowed for the formation of subatomic particles and, eventually, atoms.

Gravity then played a role in the formation of larger structures like galaxies, stars, and planets.

The theory is supported by various observations, including the cosmic microwave background radiation, which is considered a faint afterglow of the Big Bang.

While the Big Bang theory is the dominant model, some scientists explore alternative ideas, such as eternal inflation or a cyclical universe, to address unresolved questions about the universe's origins.

### 24.3 Specialty of Universe

The universe's most remarkable feature is its sheer vastness and the variety of celestial objects it contains, from tiny subatomic particles to massive galaxies.

The universe is incredibly large, with distances measured in light-years (the distance light travels in a year). Our own galaxy, the Milky Way, is vast, and there are billions of other galaxies, each containing billions of stars.

The universe is populated with a wide range of objects, including stars, planets, galaxies, nebulae, black holes, and more.

Observations show that the universe is expanding, meaning the space between galaxies is increasing.

The prevailing cosmological theory[33] suggests the universe originated from a single point and has been expanding ever since.

The universe is primarily composed of hydrogen and helium, with other elements making up a small percentage.

Despite our advancements in understanding, the universe still holds many mysteries, such as the nature of dark matter (*a mysterious substance that affects and shapes the cosmos*), etc.

While it seems complex, the universe is relatively homogeneous, meaning conditions are similar throughout, allowing scientists to make broad generalizations.

## 24.4 Galaxies

A galaxy is a system of stars, stellar remnants, interstellar gas, dust, and dark matter bound together by gravity. The word is derived from the Greek galaxias, literally 'milky', a reference to the Milky Way Galaxy that contains the Solar System.

The largest known galaxy in the universe is **IC 1101**—a supergiant elliptical galaxy located at the heart of the Abell 2029 galaxy cluster. It dwarfs even the Milky Way, with a diameter estimated to be around 5.5 million to 6 million light-years. **IC 1101** is also exceptionally luminous and contains an estimated 100 trillion stars.

**IC 1101:**
As mentioned, this galaxy is the largest known, situated within the Abell 2029 galaxy cluster.

**ESO 383–76:**
This galaxy is another contender for the title of largest, with a diameter of 1.76 million light-years.

**NGC 6872:**
A large spiral galaxy also known as the Condor Galaxy, and it is about 522,000 light-years across.

**UGC 2885:**
A giant spiral galaxy, which is one of the largest galaxies known, with a diameter of approximately 463,000 light-years.[34]

**Abell 2261-BCG:**
This brightest cluster galaxy in Abell 2261 is also among the largest, with a diameter of about 545,000 light-years.

**Hercules A:**
A radio galaxy, Hercules A is also a significant, large galaxy with a diameter of around 460,000 light-years.

**ESO 306–17:**
Another large galaxy, which has a diameter of 1.07 million light-years according to a YouTube short.

**Comet Galaxy:**
This galaxy is roughly 600,000 light-years across according to World Atlas.

**Messier 87:**
A massive elliptical galaxy, Messier 87 is known for its powerful jet of plasma.

**Phoenix Cluster (Brightest Galaxy):**
The central galaxy of the Phoenix cluster is also very large, measuring 2.2 million light-years across.

### *24.4.1 Types of Galaxies*

The four main types of galaxies include any of the systems of stars and interstellar matter that make up the universe. Many such assemblages are so enormous that they contain hundreds of billions of stars, and are spiral, elliptical, irregular, and peculiar. These classifications are based on their visual morphology, or shape.

#### *24.4.1.1 Spiral Galaxies*

These galaxies have a central bulge and spiral arms extending outwards, giving them a pinwheel shape. They often contain both young and old stars, as well as gas and dust. Our own Milky Way Galaxy is an example of a spiral galaxy.

#### *24.4.1.2 Elliptical Galaxies*

These galaxies are shaped like spheres or elongated ellipsoids and contain mostly older stars with very little gas and dust. They tend to be less organized than spiral galaxies.

#### *24.4.1.3 Irregular Galaxies*

These galaxies have no defined shape or structure. They may be the result of galactic collisions or interactions.

#### *24.4.1.4 Peculiar Galaxies*

These galaxies do not fit neatly into the other categories, and often have unusual features or structures due to past interactions or events.

Nature has provided an immensely varied array of galaxies, ranging from faint, diffuse dwarf objects to brilliant spiral-shaped giants. Virtually all galaxies appear to have been formed soon after the universe began, and they pervade space, even into the depths of the farthest reaches penetrated by powerful modern telescope.

Galaxies usually exist in clusters, some of which in turn are grouped into larger clusters that measure hundreds of millions of light-years (the distance travelled by light moving in a vacuum in the course of one year, at its accepted velocity of 299,792,458 metres per second (186,282 miles per second)). A light-year equals about $9.46073 \times 10^{12}$ km ($5.87863 \times 10^{12}$ miles), or 63,241 astronomical units. About 3.262 light-years equal one parsec (unit for expressing distances to stars and galaxies, used by professional astronomers). It represents the distance at which the radius of Earth's orbit subtends an angle of one second of arc.

Harvard and Smithsonian Center for Astrophysics say that scientists now estimate that there are over a trillion galaxies in the universe. Many follow broad classification patterns, but no two are exactly alike. The variety of galaxies we see gives us a picture of a richly differentiated universe.

The galaxies we can see with current technology, including NASA's Hubble Space Telescope, are only a small fraction of all the galaxies in the universe. This means that most of the galaxies in the universe are too dim and distant to be seen with current telescopes, and that there were many more galaxies in the early universe than there are now.

Galaxies change over hundreds of millions of years, and we cannot see a timelapse play out in our observations of the universe. However, scientists have made great strides in theoretical and computational modelling of galaxies as they change over these long timescales, enabling our understanding of galaxy evolution to constantly progress.

## 24.5 Wonders of the Universe

The universe offers countless wonders, from celestial bodies like the rings of Saturn and the Great Red Spot of Jupiter to deep-sky objects like star clusters and nebulae. Some wonders like the Orien Nebula are visible with basic telescopes, while others like black holes[35] and supernova remnants require more advanced instruments. The study of these wonders is a continuing journey, revealing the vastness and complexity of the cosmos.

## 24.6 Universe and the Earth

The universe encompasses all of space, time, matter, and energy. It is a dynamic, expanding realm containing everything that we can observe and potentially more.

Our solar system is a small part of one galaxy, the Milky Way, which is just one of billions of galaxies in the universe. The solar system consists of the sun and eight planets, including Earth, that orbit the sun.

Earth is a unique planet, the only one known to harbour life. It is a sphere, rotating on its axis (causing day and night) and orbiting the sun (causing seasons). Earth also has a moon that orbits it.

Galaxies are massive collections of stars, gas, dust, and dark matter, held together by gravity. Our galaxy, the Milky Way, is a spiral galaxy, and there are countless other galaxies, some of which are also spiral, while others are elliptical or irregular.

The universe is not static; it is constantly expanding, with galaxies moving further apart from each other.

In the vast, expanding space known as the universe, humans reside on a small, rocky planet called Earth. Our planet is part of a discrete solar system in an arm of the spiral-shaped Milky Way Galaxy. Our galaxy is only one of billions of other galaxies that exist within the universe.

According to Science Daily, about 8.7 million is the new, estimated total number of species on Earth—the most precise calculation ever offered—with 6.5 million species on land and 2.2 million in oceans. Announced by the Census of Marine Life, the figure is based on a new analytical technique. The number of species on Earth had been estimated previously at 3 million to 100 million.

# 25 Epilogue

Friends, in this book, we have tried our best to get a glimpse of the power potential of Mother Nature that creates, nurtures, and terminates all the animate and inanimate species, beginning from the one-sense plant kingdom to six-sense humans that prevail in the universe.

It is well known that all living species, beginning from the one-sense plant kingdom to six-sense humans, are made up of the five basic elements of Nature. The soil, air, water, heat, and space are the basic elements. The spectral content, executive power of these elements and ways to stimulate this power and bring it out remains as a black box. The conclusions derived by great saints and scientists, who attempted to understand this, covey that it is impossible to quantify this power. Despite this awareness, we ventured into this formidable task simply because of the command order dictated by our mind, which is the supreme commander of our body as well as the soul.

From our intense study, we realized that the power of the mind might be realized, that too with extraordinary efforts, but cannot be quantified. In addition, the moment our thought enters the path of reaching the mind, the waves emitted by the mind will immerse the soul in eternal bliss, to a speechless state. Thus, even those who worked hard and reached the foot of this insurmountable mountain cannot express their feelings to society.

Finally, a word of optimism. Let us believe that the facts, consequences, and the end results we have analysed, with the power of our mind, will raise us to the level of practising meditation, climbing all four steps of it, and enjoying the pleasures of eternal peace.

DOI: 10.4324/9781003745051-25

# Reference List


1. Puliyur Kesigan, *Tholkappiyam* (Chennai: Paari Nilayam; 2nd ed., 2012).
2. Ethirajan Rathakrishnan, *Elements of Heat Transfer* (CRC Press Inc; 1st ed., 2012).
3. Tetsuo Kobayashi, and Kagoshima-Ken, *A Guide Book for Sakurajima Volcano*, Kagoshima International Conference on Volcanoes, Kagoshima, Japan, July 19–23, 1988.
4. A. M. Mayer, and A. Poljakoff-Mayber, *Germination of Seeds* (Pergamon Press; 3rd ed., 1982).
5. Rick Stroud, *The Book of the Moon* (Doubleday, 2009).
6. Ethirajan Rathakrishnan, *Mind Power: The Sixth Sense* (Routledge, 2023).
7. Leon Golub, and Jay M. Pasachoff, *Near Star: The Surprising Science of Our Sun* (Cambridge University Press, 2014).
8. S. M. Hanasoge, and K. R. Sreenivasan, From the cover: Anomalously weak solar convection, *Proceedings of the National Academy of Sciences*, vol. 109, no. 30, pp. 11928–11932, 2012.
9. C. E. Backus, *Solar Cells* (New York: IEEE, 1976, p. 512).
10. J. E. Parrott, Choice of an equivalent black body solar temperature, *Solar Energy*, vol. 51, p. 195, 1993.
11. Linda Chaikin, *Lions of the Desert* (Multnomah Books, 1997).
12. Elizabeth Kolbert, *The Sixth Extinction: An Unnatural History* (Picador; Reprint ed., 2015).
13. The National Oceanic and Atmospheric Administration (NOAA) U.S. federal scientific and regulatory agency, Department of Commerce.
14. The International Tsunami Information Center (ITIC), established in 1965, 1845 Wasp Boulevard, Building 176, Honolulu, HI 96818, USA.
15. Kenneth Chang, The destructive power of nature, *The New York Times*, March 12, 2011.
16. Alicia Z. Klepeis, *The Science of Volcanic Eruptions (The Science of Natural Disasters)* (Cavendish Square, 2019).
17. Radcliffe G. Edmonds, III, *Myths of the Underworld Journey: Plato, Aristophanes, and the â Orphic Gold Tablets* (Cambridge University Press, 2012).
18. Peter Godfrey-Smith, *Complexity and the Function of Mind in Nature* (Cambridge University Press, 1996).
19. Elaine Landau, *Emperor Penguins: Animals of the Snow and Ice* (Enslow Pub Inc., 2010).
20. Ethirajan Rathakrishnan, *High Enthalpy Gas Dynamics* (Wiley, 2014).
21. Stephen M. Barr, *A Students Guide to Natural Science* (Wilmington, DE: Intercollegiate Studies Institute, 2006).

22. Edward Grant, *A History of Natural Philosophy: From the Ancient World to the 19th Century* (Cambridge University Press, 2007).
23. Hugh G. Gauch, *Scientific Method in Practice* (Cambridge University Press, 2003, pp. 71–73).
24. Brian W. Oglivie, *The Science of Describing: Natural History in Renaissance Europe* (University of Chicago Press, 2008).
25. *Natural History*, Princeton University WordNet. Archived from the original on March 3, 2012. Retrieved October 21, 2012.
26. Simon Fitzpatrick, Nativism, empiricism, and Ockhamâs Razo, *Erkenntnis*, vol. 80, no. 5, pp. 895–922, 2015.
27. Noam Chomsky, *The Mysteries of Nature: How Deeply Hidden? What Kind of Creatures Are We?* (Columbia University Press, 2015, pp. 81–128).
28. Heather L. Montgomery, *Unsolved Mysteries of Nature (Unsolved Mystery Files)* (Capstone Pr. Inc., 2015).
29. Michael Bennett, *Artificial Intelligence vs. Human Intelligence: Differences Explained* (Northeastern University, October 7, 2024).
30. Sarra Reiman, Is space an environment? *Space Policy*, vol. 25, no. 2, pp. 81–87, 2009.
31. Don Lincoln, *Understanding the Universe* (World Scientific Pub Co Inc., 2012).
32. Steven Weinberg, *The First Three Minutes: A Modern View of the Origin of the Universe* (Basic Books, 1993).
33. Stephen Hawking, *Universal in a Nutshell* (Bantam Press, 2001).
34. Collins Maps, *World Atlas* (Collins; 2nd ed., 2012).
35. Stephen Hawking, *Black Holes (L): The Reith Lectures* (Bantam Books, 2016).

# Index

For Product Safety Concerns and Information please contact our
EU representative GPSR@taylorandfrancis.com Taylor & Francis
Verlag GmbH, Kaufingerstraße 24, 80331 München, Germany